Synthesis, Properties and Applications of Germanium Chalcogenides

Synthesis, Properties and Applications of Germanium Chalcogenides

Editor

Stefania M. S. Privitera

MDPI • Basel • Beijing • Wuhan • Barcelona • Belgrade • Manchester • Tokyo • Cluj • Tianjin

Editor
Stefania M. S. Privitera
Institute for Microelectronics and Microsystems (IMM),
National Research Council (CNR)
Italy

Editorial Office
MDPI
St. Alban-Anlage 66
4052 Basel, Switzerland

This is a reprint of articles from the Special Issue published online in the open access journal *Nanomaterials* (ISSN 2079-4991) (available at: https://www.mdpi.com/journal/nanomaterials/special_issues/Ge_chalcogenides).

For citation purposes, cite each article independently as indicated on the article page online and as indicated below:

LastName, A.A.; LastName, B.B.; LastName, C.C. Article Title. *Journal Name* **Year**, *Volume Number*, Page Range.

ISBN 978-3-0365-5261-3 (Hbk)
ISBN 978-3-0365-5262-0 (PDF)

© 2022 by the authors. Articles in this book are Open Access and distributed under the Creative Commons Attribution (CC BY) license, which allows users to download, copy and build upon published articles, as long as the author and publisher are properly credited, which ensures maximum dissemination and a wider impact of our publications.

The book as a whole is distributed by MDPI under the terms and conditions of the Creative Commons license CC BY-NC-ND.

Contents

About the Editor vii

Stefania M. S. Privitera
Synthesis, Properties and Applications of Germanium Chalcogenides
Reprinted from: *Nanomaterials* **2022**, *12*, 2925, doi:10.3390/nano12172925 1

Marco Bertelli, Adriano Díaz Fattorini, Sara De Simone, Sabrina Calvi, Riccardo Plebani, Valentina Mussi, Fabrizio Arciprete, Raffaella Calarco and Massimo Longo
Structural and Electrical Properties of Annealed $Ge_2Sb_2Te_5$ Films Grown on Flexible Polyimide
Reprinted from: *Nanomaterials* **2022**, *12*, 2001, doi:10.3390/nano12122001 5

Minh Anh Luong, Nikolay Cherkashin, Béatrice Pecassou, Chiara Sabbione, Frédéric Mazen and Alain Claverie
Effect of Nitrogen Doping on the Crystallization Kinetics of $Ge_2Sb_2Te_5$
Reprinted from: *Nanomaterials* **2021**, *11*, 1729, doi:10.3390/nano11071729 15

Xudong Wang, Xueyang Shen, Suyang Sun and Wei Zhang
Tailoring the Structural and Optical Properties of Germanium Telluride Phase-Change Materials by Indium Incorporation
Reprinted from: *Nanomaterials* **2021**, *11*, 3029, doi:10.3390/nano11113029 31

Stefano Cecchi, Iñaki Lopez Garcia, Antonio M. Mio, Eugenio Zallo, Omar Abou El Kheir, Raffaella Calarco, Marco Bernasconi, Giuseppe Nicotra and Stefania M. S. Privitera
Crystallization and Electrical Properties of Ge-Rich GeSbTe Alloys
Reprinted from: *Nanomaterials* **2022**, *12*, 631, doi:10.3390/nano12040631 43

Adriano Díaz Fattorini, Caroline Chèze, Iñaki López García, Christian Petrucci, Marco Bertelli, Flavia Righi Riva, Simone Prili, Stefania M. S. Privitera, Marzia Buscema, Antonella Sciuto, Salvatore Di Franco, Giuseppe D'Arrigo, Massimo Longo, Sara De Simone, Valentina Mussi, Ernesto Placidi, Marie-Claire Cyrille, Nguyet-Phuong Tran, Raffaella Calarco and Fabrizio Arciprete
Growth, Electronic and Electrical Characterization of Ge-Rich Ge–Sb–Te Alloy
Reprinted from: *Nanomaterials* **2022**, *12*, 1340, doi:10.3390/nano12081340 57

Daniel Tadesse Yimam, A. J. T. Van Der Ree, Omar Abou El Kheir, Jamo Momand, Majid Ahmadi, George Palasantzas, Marco Bernasconi and Bart J. Kooi
Phase Separation in Ge-Rich GeSbTe at Different Length Scales: Melt-Quenched Bulk versus Annealed Thin Films
Reprinted from: *Nanomaterials* **2022**, *12*, 1717, doi:10.3390/nano12101717 69

Arun Kumar, Raimondo Cecchini, Claudia Wiemer, Valentina Mussi, Sara De Simone, Raffaella Calarco, Mario Scuderi, Giuseppe Nicotra and Massimo Longo
Phase Change Ge-Rich Ge–Sb–Te/Sb_2Te_3 Core-Shell Nanowires by Metal Organic Chemical Vapor Deposition
Reprinted from: *Nanomaterials* **2021**, *11*, 3358, doi:10.3390/nano11123358 83

Omar Abou El Kheir and Marco Bernasconi
High-Throughput Calculations on the Decomposition Reactions of Off-Stoichiometry GeSbTe Alloysfor Embedded Memories
Reprinted from: *Nanomaterials* **2021**, *11*, 2382, doi:10.3390/2382 nano11092382 93

Valeria Bragaglia, Vara Prasad Jonnalagadda, Marilyne Sousa, Syed Ghazi Sarwat, Benedikt Kersting and Abu Sebastian
Structural Assessment of Interfaces in Projected Phase-Change Memory
Reprinted from: *Nanomaterials* **2022**, *12*, 1702, doi:10.3390/nano12101702 . . . . **107**

Caroline Chèze, Flavia Righi Riva, Giulia Di Bella, Ernesto Placidi, Simone Prili, Marco Bertelli, Adriano Diaz Fattorini, Massimo Longo, Raffaella Calarco, Marco Bernasconi, Omar Abou El Kheir and Fabrizio Arciprete
Interface Formation during the Growth of Phase Change Material Heterostructures Based on Ge-Rich Ge-Sb-Te Alloys
Reprinted from: *Nanomaterials* **2022**, *12*, 1007, doi:10.3390/nano12061007 . . . . **123**

Arun Kumar, Seyed Ariana Mirshokraee, Alessio Lamperti, Matteo Cantoni, Massimo Longo and Claudia Wiemer
Interface Analysis of MOCVD Grown GeTe/Sb_2Te_3 and Ge-Rich Ge-Sb-Te/Sb_2Te_3 Core-Shell Nanowires
Reprinted from: *Nanomaterials* **2022**, *12*, 1623, doi:10.3390/nano12101623 . . . . **135**

About the Editor

Stefania M. S. Privitera

Stefania M. S. Privitera received her B.S. degree (cum laude) in 1998 and her Ph.D. degree in Physics from the University of Catania, Italy, in 2002. From 2002 to 2004, she was a post-doc researcher at the University of Catania, working on phase change materials for non-volatile memories. In 2004, she joined STMicroelectronics at the R&D Department, working on the design and integration of electronic devices. Since December 2011, she is a Staff Research Scientist of the National Research Council (CNR) at the Institute for Microelectronics and Microsystems (IMM), Catania. Her research interests are in the field of the development of semiconductor devices and of electrical and structural characterization of electronic materials, with focus on phase change materials and on novel catalysts for solar fuels production. She is author of six US-issued patents, about 76 scientific and technical papers published in international refereed journals, and more than 40 technical reports, and she has contributed to two book chapters. She has presented oral and poster contributions at numerous international conferences/workshops, including six invited presentations at international conferences.

She is the Guest Editor of the Special Issue "Materials and Devices for Solar to Hydrogen Energy Conversion", Energies, MDPI, and of the Special Issue "Synthesis, Properties and Applications of Germanium Chalcogenides", *Nanomaterials*, MDPI.

She is also chair of the symposium on Phase Change Materials at the 2017 MRS Spring Meeting, Phoenix (US), and at the 2018 MRS Spring Meeting.

Editorial

Synthesis, Properties and Applications of Germanium Chalcogenides

Stefania M. S. Privitera

Institute for Microelectronic and Microsystems (IMM), National Research Council (CNR),
Zona Industriale Ottava Strada 5, 95121 Catania, Italy; stefania.privitera@imm.cnr.it

Citation: Privitera, S.M.S. Synthesis, Properties and Applications of Germanium Chalcogenides. *Nanomaterials* **2022**, *12*, 2925. https://doi.org/10.3390/nano12172925

Received: 9 August 2022
Accepted: 23 August 2022
Published: 25 August 2022

Publisher's Note: MDPI stays neutral with regard to jurisdictional claims in published maps and institutional affiliations.

Copyright: © 2022 by the author. Licensee MDPI, Basel, Switzerland. This article is an open access article distributed under the terms and conditions of the Creative Commons Attribution (CC BY) license (https://creativecommons.org/licenses/by/4.0/).

Germanium (Ge) chalcogenides are characterized by unique properties which make these materials interesting for a very wide range of applications, from phase change memories to ovonic threshold switches, from photonics to thermoelectric and photovoltaic devices. In many cases, physical properties can be finely tuned by doping or by changing the Ge amount, which can thus play a key role in determining the applications, performance, and even the reliability of the devices. In this Special Issue, we include 11 articles, mainly focusing on applications of Ge chalcogenides for nonvolatile memories.

Most of the papers were produced with funding received from the European Union's Horizon 2020 Research and Innovation program under grant agreement n. 824957 for the project "BeforeHand: Boosting Performance of phase change Devices by Hetero- and Nanostructure Material Design".

Two contributions [1,2] are related to the prototypical $Ge_2Sb_2Te_5$ compound, which is a widely studied composition and is already integrated in many devices such as optical and electronic memories. In [1], M. Bertelli at al. report on the structural and electrical properties of GST225 grown on polyimide, a flexible substrate whose use could enable novel applications in the market of electronics, for example, flexible nonvolatile memories for the IoT, or smart sensors for food and drug monitoring. The paper reports information about the layer evolution during amorphous-to-cubic and cubic-to-trigonal transitions, and the related electrical contrast.

In [2], M. A. Luong et al. investigate the atomistic mechanisms related to nitrogen doping, which is known to improve some key characteristics of the materials, such as the amorphous stability and the resistance drift. These effects are ascribed to the increased viscosity of the N-doped amorphous state and to the reduced diffusivity resulting from the formation of N-Ge bonds, demonstrating that the origin of the effect of N on crystallization is attributed to the ability of N to bind to Ge in the amorphous and crystalline phases and to unbind and rebind with Ge along the diffusion path during annealing.

Another approach to improve the thermal stability of the amorphous phase is presented in [3], where starting from the GeTe alloy, X. Wang and coauthors incorporate indium, obtaining three typical compositions in the InTe-GeTe tie line, and propose a chemical composition with both improved thermal stability and sizable optical contrast for photonic applications.

Ge-rich GeSbTe (GST) alloys are currently explored for embedded memory applications, with the aim to increase the crystallization temperature, therefore improving the amorphous phase stability. However, deposited homogenous alloys are thermodynamically unstable and undergo phase separation upon annealing.

Five articles of this Special Issue focus on Ge-rich GST alloys, exploring their electronic and electrical properties [4–7] as well as decomposition pathways, including from a theoretical point of view [8].

In [4], S. Cecchi et al. identify some possible routes to limit Ge segregation, investigating Ge-GST compositions deposited by molecular beam epitaxy in the amorphous phase with low or high (>40%) amounts of Ge. Electrical resistance and phase formation are studied upon annealing up to 300 °C.

In [5], A. Diaz Fattorini and coauthors deposit Ge-rich GST with a composition of $Ge_{29}Sb_{20}Te_{28}$ via physical vapour deposition (PVD). They study the electronic properties and phase formation and report the electrical characterization of a single memory cell, showing the possibility to enhance the thermal stability up to 230 °C while maintaining a fair alignment of electrical parameters with the current state of the art of conventional GST alloys.

The contribution of D. Tadesse Yimam et al. [6] investigates the phase separation of GST523 into multiple phases in melt quenched bulk and annealed thin films, identifying the formation of GST123 and GST324 alloys in all length scales.

The alloy compositions and the observed phase separation pathways reported in [4,6] agree to a large extent with the theoretical results from the density functional theory calculations, as presented in [8], where O. Abou El Kheir and M. Bernasconi perform high-throughput calculations to uncover the most favorable decomposition pathways of Ge-rich GST alloys. They also construct a map of decomposition propensity, suggesting a possible strategy to minimize phase separation while still maintaining a high crystallization temperature.

In [7], A. Kumar and coauthors investigate the effect of Ge-rich GST in nanowires self-assembled through the vapor–liquid–solid mechanism. Both Ge-rich GST core and Ge-rich GST/Sb_2Te_3 core shells are extensively characterized with several techniques to analyze the surface morphology, crystalline structure, vibrational properties and elemental composition.

Other tree contributions [9–11] are focused on the effect of the interfaces, since in nanomaterials, element interdiffusion at the interfaces represents a crucial factor.

In [9], V. Bragaglia et al. investigate this aspect in projected phase change memories, in which the storage mechanism is decoupled from the information retrieval process via a projection liner. The interface resistance between the phase change chalcogenide material and the projection liner is an important parameter, and therefore a metrology framework is established to assess the quality of the interfaces through X-ray reflectivity, X-ray diffraction, and transmission electron microscopy.

As another important case in which interfaces play a significant role, article [10] by C. Chèze and coauthors reports the full characterization of the electronic properties of double-layered heterostructures made by Ge-rich GST deposited by PVD on Sb_2Te_3 and on $Ge_2Sb_2Te_5$. Information on interdiffusion and on the evolution of the composition across the interface was obtained; it was found that, in both heterostructures, the final composition was GST212, which is a thermodynamically favorable off-stoichiometry alloy in the Sb-GeTe pseudo-binary line.

The interdiffusion at the interface of core–shell nanowires with a Sb_2Te_3 shell over GeTe and a Ge-rich GST core is studied in [11] by Kumar et al. by examining the morphological and structural characteristics. No elemental interdiffusion between core and shell is revealed, suggesting that their structural phases can change independently based on alloy compositions, thus demonstrating a straightforward method to provide core–shell nanowire heterostructures formed by two-phase chalcogenide materials with different crystallization temperatures and switching speeds.

Funding: This research received no external funding.

Conflicts of Interest: There are not conflict of Interest.

References

1. Bertelli, M.; Fattorini, A.D.; De Simone, S.; Calvi, S.; Plebani, R.; Mussi, V.; Arciprete, F.; Calarco, R.; Longo, M. Structural and Electrical Properties of Annealed $Ge_2Sb_2Te_5$ Films Grown on Flexible Polyimide. *Nanomaterials* **2022**, *12*, 2001. [CrossRef] [PubMed]
2. Luong, M.; Cherkashin, N.; Pecassou, B.; Sabbione, C.; Mazen, F.; Claverie, A. Effect of Nitrogen Doping on the Crystallization Kinetics of $Ge_2Sb_2Te_5$. *Nanomaterials* **2021**, *11*, 1729. [CrossRef] [PubMed]
3. Wang, X.; Shen, X.; Sun, S.; Zhang, W. Tailoring the Structural and Optical Properties of Germanium Telluride Phase-Change Materials by Indium Incorporation. *Nanomaterials* **2021**, *11*, 3029. [CrossRef] [PubMed]

4. Cecchi, S.; Garcia, I.L.; Mio, A.M.; Zallo, E.; El Kheir, O.A.; Calarco, R.; Bernasconi, M.; Nicotra, G.; Privitera, S.M.S. Crystallization and Electrical Properties of Ge-Rich GeSbTe Alloys. *Nanomaterials* **2022**, *12*, 631. [CrossRef] [PubMed]
5. Fattorini, A.D.; Chèze, C.; García, I.L.; Petrucci, C.; Bertelli, M.; Riva, F.R.; Prili, S.; Privitera, S.M.S.; Buscema, M.; Sciuto, A.; et al. Growth, Electronic and Electrical Characterization of Ge-Rich Ge–Sb–Te Alloy. *Nanomaterials* **2022**, *12*, 1340. [CrossRef] [PubMed]
6. Yimam, D.T.; Van Der Ree, A.J.T.; El Kheir, O.A.; Momand, J.; Ahmadi, M.; Palasantzas, G.; Bernasconi, M.; Kooi, B.J. Phase Separation in Ge-Rich GeSbTe at Different Length Scales: Melt-Quenched Bulk versus Annealed Thin Films. *Nanomaterials* **2022**, *12*, 1717. [CrossRef] [PubMed]
7. Kumar, A.; Cecchini, R.; Wiemer, C.; Mussi, V.; De Simone, S.; Calarco, R.; Scuderi, M.; Nicotra, G.; Longo, M. Phase Change Ge-Rich Ge–Sb–Te/Sb2Te3 Core-Shell Nanowires by Metal Organic Chemical Vapor Deposition. *Nanomaterials* **2021**, *11*, 3358. [CrossRef] [PubMed]
8. El Kheir, O.A.; Bernasconi, M. High-Throughput Calculations on the Decomposition Reactions of Off-Stoichiometry GeSbTe Alloys for Embedded Memories. *Nanomaterials* **2021**, *11*, 2382. [CrossRef] [PubMed]
9. Bragaglia, V.; Jonnalagadda, V.P.; Sousa, M.; Sarwat, S.G.; Kersting, B.; Sebastian, A. Structural Assessment of Interfaces in Projected Phase-Change Memory. *Nanomaterials* **2022**, *12*, 1702. [CrossRef] [PubMed]
10. Chèze, C.; Riva, F.R.; Di Bella, G.; Placidi, E.; Prili, S.; Bertelli, M.; Fattorini, A.D.; Longo, M.; Calarco, R.; Bernasconi, M.; et al. Interface Formation during the Growth of Phase Change Material Heterostructures Based on Ge-Rich Ge-Sb-Te Alloys. *Nanomaterials* **2022**, *12*, 1007. [CrossRef] [PubMed]
11. Kumar, A.; Mirshokraee, S.A.; Lamperti, A.; Cantoni, M.; Longo, M.; Wiemer, C. Interface Analysis of MOCVD Grown GeTe/Sb$_2$Te$_3$ and Ge-Rich Ge-Sb Te/Sb$_2$Te$_3$ Core-Shell Nanowires. *Nanomaterials* **2022**, *12*, 1623. [CrossRef] [PubMed]

Article

Structural and Electrical Properties of Annealed Ge₂Sb₂Te₅ Films Grown on Flexible Polyimide

Marco Bertelli [1], Adriano Díaz Fattorini [1], Sara De Simone [1], Sabrina Calvi [2,1], Riccardo Plebani [2], Valentina Mussi [1], Fabrizio Arciprete [2], Raffaella Calarco [1,*] and Massimo Longo [1]

[1] Institute for Microelectronics and Microsystems (IMM), Consiglio Nazionale delle Ricerche (CNR), Via del Fosso del Cavaliere 100, 00133 Rome, Italy; marco.bertelli@artov.imm.cnr.it (M.B.); adriano.diazfattorini@artov.imm.cnr.it (A.D.F.); sara.desimone@artov.imm.cnr.it (S.D.S.); valentina.mussi@artov.imm.cnr.it (V.M.); massimo.longo@artov.imm.cnr.it (M.L.)

[2] Department of Physics, University of "Tor Vergata", Via della Ricerca Scientifica 1, 00133 Rome, Italy; sabrina.calvi@roma2.infn.it (S.C.); plebaniriccardo@gmail.com (R.P.); fabrizio.arciprete@roma2.infn.it (F.A.)

* Correspondence: raffaella.calarco@artov.imm.cnr.it

Abstract: The morphological, structural, and electrical properties of as-grown and annealed Ge₂Sb₂Te₅ (GST) layers, deposited by RF-sputtering on flexible polyimide, were studied by means of optical microscopy, atomic force microscopy, X-ray diffraction, Raman spectroscopy, and electrical characterization. The X-ray diffraction annealing experiments showed the structural transformation of GST layers from the as-grown amorphous state into their crystalline cubic and trigonal phases. The onset of crystallization of the GST films was inferred at about 140 °C. The vibrational properties of the crystalline GST layers were investigated via Raman spectroscopy with mode assignment in agreement with previous works on GST films grown on rigid substrates. The electrical characterization revealed a good homogeneity of the amorphous and crystalline trigonal GST with an electrical resistance contrast of 8×10^6.

Keywords: PCM; Ge₂Sb₂Te₅; sputtering; flexible substrates; crystallization; electrical properties

1. Introduction

Tremendous advances in interconnected technologies are already revolutionizing our lives by integrating networks of physical devices ('things') that are capable of capturing, processing, and sharing data over the internet (Internet of Things (IoT)). Neuromorphic computing based on innovative concepts inspired by the functioning or organization of the brain's data processing chain provides answers to such challenges [1].

Revolutionary advances in computing architecture require different methods of managing computing memory and the development of alternative technologies for its implementation. This can be realized in practice by exploiting the physical properties of devices that show a certain degree of time nonlocality (memory) in their response functions, such as memristors [2].

Synaptic connections have been realized with different types of memristors [3–5], among others, with phase change memories (PCMs) [6,7], being extensively used in the semiconductor industry [8]. PCMs are based on chalcogenide alloys that are able to reversibly change their resistance upon the application of proper electrical stimuli. The resistance change exploits the different resistivities of two distinct structural solid-state phases (i.e., crystalline and amorphous), while the switching mechanism is thermally induced through Joule heating [9].

The deposition of PCMs on flexible substrates may have a huge impact on the market of electronic applications, for example, flexible nonvolatile memories for the IoT [10] or smart sensors for food and drug monitoring [11].

Polyimide (PI) is one of the most interesting candidates as a flexible substrate for PCM deposition because it is insulating (electrical resistivity $\rho = 0.1–4 \times 10^{19}$ Ωcm), lightweight (density $d = 0.005–1.88$ g/cm^3), and resistant to heat (in air maximum operating temperature $T_{max} = 200–395$ °C) [12]. PI is already being used in the electronic industry to realize flexible printed circuits [13]. Moreover, the possibility to produce transparent and/or biocompatible PI films [14,15] makes them extremely attractive for wearable smart devices and biosensors.

In previous works, we have studied the synthesis and the crystallization of single Ge-Sb-Te films with variable Ge content [16], as well as heterostructures formed by planar layers of the Ge-Sb-Te system [17]. Moreover, the self-assembly and structural characterization of core-shell nanowires of the Ge-Sb-Te system was investigated [18].

In the present work, amorphous GST films were deposited by RF-sputtering on flexible PI. The layers were studied in terms of their morphological, structural, and electrical properties prior to and after crystallization using dedicated thermal annealing during structural investigation. While a few studies have already examined the growth of GST layers on flexible substrates (i.e., mica [19], polycarbonate [20], and PI [21–26]), we could not find a systematic study on GST/PI structures reporting the morphological, structural, vibrational, and electrical properties of as-deposited (amorphous) and annealed (crystalline) thin ($\leq$150 nm) GST layers. We obtained information about the layer evolution during the amorphous-to-cubic and cubic-to-trigonal transitions and the related electrical contrast. This is important for the functioning of PCM devices based on a reversible switch between material phases and the association of binary encoding to the consequent resistivity changes. Therefore, the implementation of such stacks in flexible memristive devices should provide further advantages in terms of memristor performance (e.g., reduction in the switching energy and higher storage density).

2. Materials and Methods

2.1. Samples Growth

The GST layers were deposited by RF-sputtering in a custom-made high-vacuum chamber system (IONVAC PROCESS srl, Pomezia, Italy) equipped with a planetary system for deposition and four confocal targets. GST films (thickness ~140 nm) were grown at room temperature (RT) on flexible PI (DuPont™ Kapton® HN, thickness = 125 µm, Wilmington, DE, USA), which was coated with a protective silicon-nitride layer (thickness ~200 nm). The GST target was provided by Robeko GmbH & Co., KG (Mehlingen, Germany), with a nominal composition of $Ge_2Sb_2Te_5$ with 99.99% purity. The sputtering forward power was 50 W, and the flow of 0.04 L/min of pure argon was delivered to the growth chamber. During the deposition, the GST growth rate was monitored with an STM-2 rate monitor (INFICON Holding AG, Bad Ragaz, Switzerland), and the chamber pressure was in the range of high 10^{-2} mbar. The resulting GST deposition rate was around 0.72 nm/s.

2.2. Morphology Investigation

The morphological characterization of the GST layers was performed ex situ by means of an optical microscope (CARL ZEISS AG, Oberkochen, Germany) and an atomic force microscope (AFM) (PARK SYSTEMS Corp., Suwon, Korea) operating in noncontact mode.

2.3. X-ray Fluorescence (XRF)

Energy dispersive X-ray fluorescence (XRF) measurements were performed ex situ using a RIGAKU Nex DE VS spectrometer (APPLIED RIGAKU TECHNOLOGIES INC., Austin, TX, USA) equipped with a 60 kV X-ray tube and a silicon drift detector.

2.4. X-ray Diffraction (XRD)

XRD measurements were performed ex situ by a D8 Discover diffractometer (BRUKER, Billerica, MA, USA) equipped with a Cu X-ray source (Cu-Kα_1 radiation $\lambda = 1.5406$ Å, 40 kV and 40 mA) and a DHS1100 dome-type heating stage (ANTON PAAR, Graz, Austria) for temperature measurements in a N_2 atmosphere. Grazing incidence diffraction (GID) (ω–2θ)

scans were acquired at different temperatures (T = 30–300 °C) during sample annealing. The average grain size of the grown GST layers was evaluated for selected diffraction peaks using the Scherrer equation [27]. The acquisition parameters were: incidence angle: 0.5°; 2θ steps: 0.08°; acquisition time: 25 min 17 s (30 °C), 25 min 22 s (130 °C), 25 min 22 s (140 °C), 1 h 16 min 7 s (200 °C), 1 h 17 min 7 s (300 °C), 1 h 29 min 59 s (PI substrate).

2.5. Raman

Raman spectra were acquired ex situ by means of a DXR2xi Raman imaging microscope (THERMOFISCHER, Waltham, MA, USA) equipped with a 532 nm laser source and a 50× objective. The Raman data acquisition was performed at RT in back-scattering geometry by using a 4 mW laser power at the sample surface.

2.6. Electrical Measurements

To characterize the GST electrical properties, the layer resistance R was measured in a probe station equipped with a model 2440 and two model 236 source meters (KEITHLEY, TEKTRONIX INC., Beaverton, OR, USA) The four-point collinear probe technique was used, and the volume resistivity was calculated as $\rho = \frac{\pi}{\ln 2} \times R \times t \times R1$, where t is the film thickness, and $R1$ is the Haldor Topsøe correction factor for thin samples of finite rectangular shape.

3. Results and Discussion

Initially, we observed that the 140 nm thick GST layer (reasonable thickness for device realization, see explanation below), deposited on a 1 × 1 cm PI substrate, appeared uniform and continuous on a large scale (inset of Figure 1a). Considering a standard mushroom-type cell, the thickness of chalcogenide layers for PCM cells ranges from a few tens of nanometers to about 200 nm, depending on the heater size. High scalability is not expected for flexible substrates, for which the devices are on larger areas, rather than on rigid substrates; hence, thicker layers of the phase change materials are involved. According to the thin-film regime, thinner layers generally ensure better growth; nevertheless, thicker layers are preferable to observe possible defects such as exfoliation. Therefore, a thickness of 140 nm was selected as a good compromise. The presence of GST in our films was clearly distinguishable by the naked eye for their brown color, in contrast with the orange, transparent PI substrate.

The optical microscope data (Figure 1a) showed, on a smaller scale, that the as-grown GST layer was continuous and transparent. Even at a smaller scale, as presented in AFM observations (Figure 1b,c), the GST film surface was fairly uniform: few protrusions, but no scratches or deeps were found. The optical microscope image of the GST film, after annealing at 300 °C for 77 min (Figure 1d), revealed a different surface morphology, with the appearance of tiny spots, which will be subject to further investigation. The AFM images on 5 × 5 μm areas showed a surface roughness RMS value of 2.3 nm and 3.5 nm for the as-grown (Figure 1b) and the annealed GST layer (Figure 1e), respectively. Nevertheless, both the as-grown (Figure 1c) and annealed (Figure 1f) layers showed, on smaller areas of 1 × 1 μm, the same value of the surface roughness, namely RMS = 1.6 nm.

The XRF measurements confirmed that the composition of the grown GST layer was (Ge:Sb:Te = 2.01:1.97:5 within the 5% error), effectively reproducing the nominal composition of the sputtered target.

Figure 2 shows the GID X-ray (ω–2θ) scans of the PI/GST layers during annealing under N_2 atmosphere (T = 30–300 °C). The XRD diffractograms are displayed in the 2θ range of 27.5–80° to highlight the peaks of crystalline GST grains; the two broad peaks that characterized the clean PI substrate at ~21.8° and ~26° (inset of Figure 2a) were thus excluded.

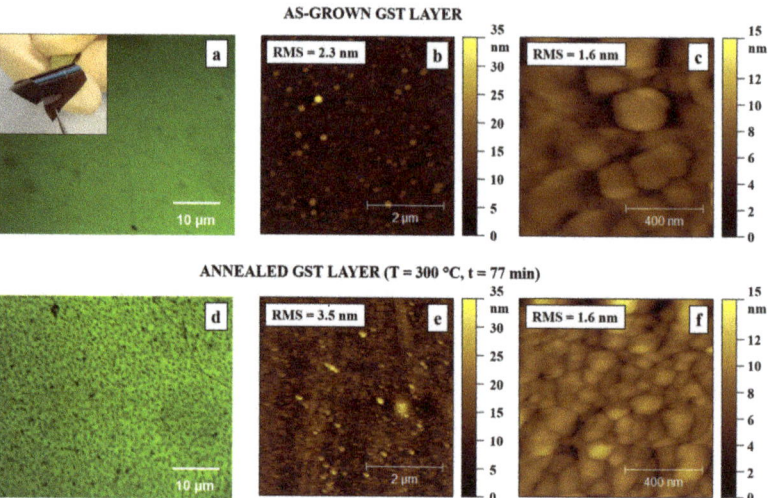

Figure 1. Ge$_2$Sb$_2$Te$_5$ (GST) layers on polyimide (PI) substrate. As-grown: (**a**) optical microscope, (inset) photograph (1 × 1 cm) sample, (**b**) atomic force microscope (AFM) (5 × 5 µm), (**c**) AFM (1 × 1 µm). After annealing at 300 °C for 77 min: (**d**) optical microscope, (**e**) AFM (5 × 5 µm), and (**f**) AFM (1 × 1 µm).

The onset of the transition from amorphous (a-)GST to crystalline GST occurred when the temperature increased from 130 to 140 °C (Figure 2a): the peaks at 42.4°, 52.5°, and 69.6° were identified as the (220), (222), and (420) Bragg reflections of GST in the cubic (c-)GST crystalline structure, respectively. The XRD peak identification was performed by comparing the measured GID X-ray scans with those of previous experimental works [17,28] and diffractograms that were calculated with the cross-platform program VESTA [29–31]. Figure 2b shows the (ω–2θ) scan at T = 200 °C after annealing at T = 200 °C for 76 min: the peaks at 29.5°, 42.6°, 52.7°, 61.6°, 69.8°, and 77.7° were identified as the (200), (220), (222), (400), (420), and (422) Bragg reflections of c-GST, respectively. A comparison of the XRD peaks revealed that the crystallization of the amorphous GST had a preferential orientation along the [220] direction. The positions of the GST reflections are in very good agreement with the expected positions; therefore, strain was not present in our sample. The GST crystallization occurred through the polycrystalline relaxed grains.

The 2θ positions of the c-GST(220), c-GST(222) and c-GST (420) peaks shifted by 0.2° to higher values for the layer annealed at T = 200 °C for 77 min in comparison with the film annealed at T = 140 °C for twenty-five minutes. The XRD peak shifting toward larger 2θ values, with increasing the annealing temperature, was already observed by Lu et al. only for the c-GST(200) XRD reflection of 4 µm thick GST layers deposited by pulsed laser deposition on PI substrates [15]. The peak displacement was due to thermal lattice expansion, as the XRD graphs were acquired at different temperatures during annealing.

Figure 2c shows the (ω–2θ) scan at T = 300 °C after the annealing of a GST layer at T = 300 °C for 77 min: the peaks at 28.6°, 39.1°, 42.6°, 43.7°, 51.8°, and 75.4° were assigned to the (10.3), (10.6), (11.0), (11.2), (20.3), and (21.6) Bragg reflections of trigonal (t-)GST, respectively. The XRD data indicated that, after such annealing, the crystallization of a-GST into the t-GST crystalline structure had occurred.

The average crystalline grain size was 8.7 and 37.8 nm for the c-GST(200) and the t-GST(10.3) diffraction peak, respectively. This result is in agreement with the trend reported by Lu et al., who observed a linear increase in the average grain size from the c-GST(200) XRD reflection as a function of the annealing temperature [24]. This is an indication of grain growth during the ordering of the material upon annealing.

The PI substrate allows for an excellent thermal stability in the considered annealing temperatures, as it is well-known in the literature [32], implying no evident degradation of the GST layer.

In Figure 3 the Raman spectra of c-GST and t-GST on PI are presented in a range from 75 to 225 cm^{-1}. The PI substrate did not contribute to the Raman spectra in the range of interest (not shown here).

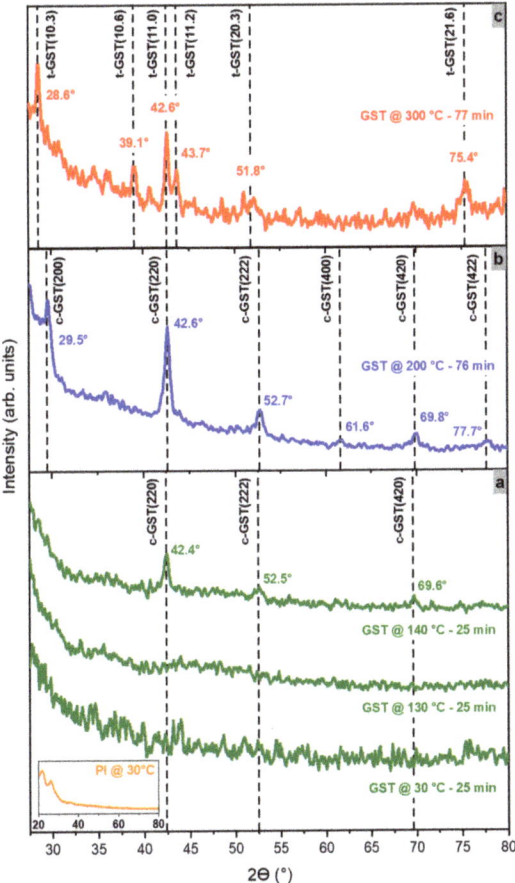

Figure 2. X-ray grazing incidence diffraction (GID) (ω–2θ) scans on PI/GST sample for (**a**) annealing experiment at T = 30, 130, and 140 °C (each temperature increase step was t = 25 min); (**b**) at T = 200 °C for t = 76 min; (**c**) at T = 300 °C for t = 77 min; (inset) of (**a**) X-ray GID (ω–2θ) scan on PI substrate at T = 30 °C acquired for t = 25 min. Dotted black lines indicate the positions of the experimental X-ray GID (ω–2θ) scan peaks.

The spectrum of the c-GST layer shows three broad peaks centered at about 118, 138, and 159 cm^{-1}. The peaks at 118 and 159 cm^{-1} were assigned to the E_g and A_{1g} modes of hexagonal Sb$_2$Te$_3$, respectively [33,34]. The peak at 138 cm^{-1} could be attributed to the A$_1$ mode of corner-sharing GeTe$_{4-n}$Ge$_n$ (*n* = 1,2) tetrahedra [33] or, alternatively, to Te segregation in the sample [35]. The spectrum of the t-GST layer shows two main broad bands located at about 109 and 164 cm^{-1}, which could be ascribed to the E- and A-type vibration modes, respectively [34].

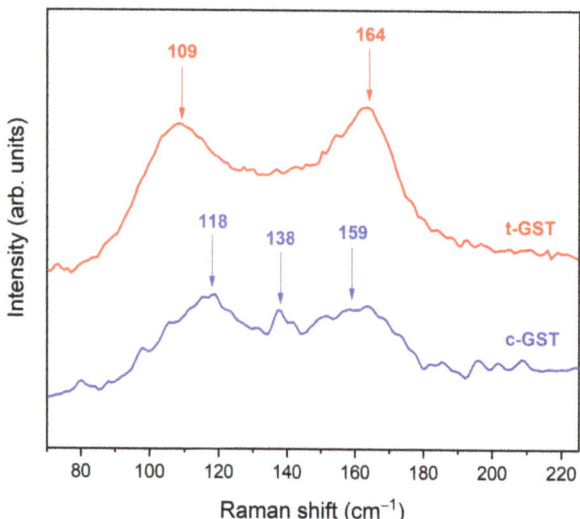

Figure 3. Raman spectra at room temperature (RT) of PI/GST sample after annealing at T = 200 °C for t = 76 min (blue curve) and after annealing at T = 300 °C for t = 77 min (red curve).

The acquisition of the Raman spectra on PI substrates was not obvious because the relatively low thermal conductivity of the PI, which is in the range of 0.2 Wm^{-1} K^{-1} [36], is considered to be one of the major challenges for avoiding thermal failure in electronic devices. This feature limits the PI thermal management, especially in terms of heat dissipation; therefore, it must be taken into account in its response to laser heating. Regardless of whether infrared (wavelength 780 nm) or green (532 nm) laser light was used as the excitation source, we were not able to record a reliable Raman spectrum from the a-GST. Even at the lowest available power (0.1 mW) and with a very low exposure time (0.5 ms), we observed an immediate process of GST crystallization, which was highlighted by the formation of brighter spots with higher reflectivity on the sample surface. Therefore, we could not exclude that the observation of the peak at 138 cm^{-1} in the c-GST Raman spectrum might have been due to possible segregation of Te related to a modification induced by the laser light, causing local heating of the substrate [37].

Several factors may affect the PCM's electrical properties, such as film thickness, deposition method, and growth parameters. The electrical characteristics of the film are of interest as a direct indication for exploitation in memory devices of as-grown layers. A relevant contrast between the electrical resistance in the amorphous and crystalline states allows for a large on/off ratio in PCM cells. Hence, the amorphous (pristine as-deposited) and crystalline (after annealing) GST films on PI were electrically characterized at RT. In Figure 4, we report the linear dependence of the resistance R as a function of the contact distance L for amorphous and crystalline PI/GST.

The data showed the good homogeneity of the GST layers, which are thus compatible with large-area scalability. This confirmed the morphology results obtained by optical and AFM microscopy. For the crystalline and amorphous films, we measured a sheet resistance of $R_{sq, cryst}$ = (332 ± 30) Ω/sq and $R_{sq, am}$ = (2.6 ± 1.3) GΩ/sq, and a resistivity of ρ_{cryst} = (4.6 ± 0.4) mΩ·cm and ρ_{am} = (36 ± 18) kΩ·cm respectively. These values are in line with those of state-of-the-art GST films grown on rigid substrates [38–41], also deposited by other techniques such as atomic layer deposition [42] or molecular beam epitaxy [43]. The phase reached by the PCM in a cell, both for the set and reset states, is strongly dependent on the configuration, contacts, scaling level, and operating conditions. Therefore, in this study, fully amorphous and trigonal crystalline phases were considered as the most suitable to ensure a clear reference and an easy comparison with the values in the literature. The resistance ratio was separated into six orders of magnitude, namely $R_{am}/R_{cryst} \approx 8 \times 10^6$, which confirmed the excellent electrical features of the material deposited on the PI substrates. This is promising, because a key requirement of a PCM memory is that the programming window is large enough to avoid the possible ambiguity of data encoding and reduce resistance drift effects. In practical PCM devices, the resistance contrast is usually reduced by the contact and heater resistances; furthermore, the melt-quenched (thus amorphized) material has a lower resistivity with respect to the as-deposited amorphous phase. However, the large resistance ratio measured in our GST layers is in line with the values reported for materials suitable for PCM memories (around five orders of magnitude, [44]); thus, it is expected to allow for a large on/off ratio in flexible PCM memristors. Moreover, in a PCM cell, the high resistance contrast enables a gradual crystallization of an amorphous region by the application of repeated pulses. A controllable fractional variation of the resistance extends the possible applications to a multilevel domain. Gradual resistive switching is a key concept for analogue computing as it enables logic summation within a single PCM cell.

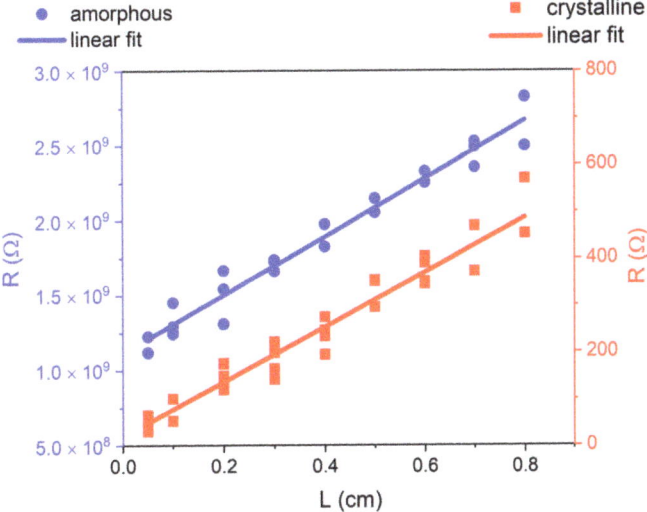

Figure 4. Resistance R at RT of the amorphous (as-deposited) and crystalline (after annealing at 300 °C for 77 min) GST films on PI, as a function of the contact distance L.

4. Conclusions

We deposited amorphous, 140 nm thick, $Ge_2Sb_2Te_5$ films by RF-sputtering on flexible polyimide. The behavior of the films upon thermal annealing was studied during the amorphous-to-cubic and cubic-to-trigonal transitions. Both the as-grown and annealed layers were uniform and continuous with surface roughness RMS of 1.6 nm on 1×1 µm areas. GST layers of 140 nm can be grown in a fully amorphous phase on polyimide and the crystallization onset occurs at about 140 °C. Upon annealing, the transition from the amorphous to the cubic and trigonal phase was observed, as also confirmed by subsequent Raman characterization. Finally, the electrical characterization showed good homogeneity of the amorphous and crystalline trigonal GST layers, exhibiting a resistance ratio of $R_{am}/R_{cryst} \approx 8 \times 10^6$. This is a remarkable result in view of applications of PCM memristors for flexible smart devices devoted to IoT applications such as wearable and biocompatible sensors.

Author Contributions: Conceptualization, R.C.; methodology, R.C., M.L. and M.B.; investigation, M.B., A.D.F., S.D.S., S.C., R.P. and V.M.; resources, R.C. and M.L.; data curation, M.B., S.D.S., S.C., V.M., R.C. and F.A.; writing—original draft preparation, M.B., M.L. and R.C.; writing—review and editing, M.B., M.L., R.C. and F.A.; funding acquisition, R.C. and M.L. All authors have read and agreed to the published version of the manuscript.

Funding: This project has received funding from the European Union's Horizon 2020 research and innovation program, under grant agreement No. 824957 ("BeforeHand": Boosting Performance of Phase Change Devices by Hetero- and Nanostructure Material Design). S.C. gratefully acknowledges the program "FSE-REACT EU—PON Research and Innovation 2014–2020", by the Italian Ministry of University and Research.

Institutional Review Board Statement: Not applicable.

Informed Consent Statement: Not applicable.

Data Availability Statement: Not applicable.

Acknowledgments: The authors acknowledge Simone Prili for the elaboration of VESTA calculated XRD diffractograms, F. Maita for technical support on the electrical characterization, and M. Scagliotti for technical support on the AFM measurements.

Conflicts of Interest: The authors declare no conflict of interest.

References

1. Marković, D.; Mizrahi, A.; Querlioz, D.; Grollier, J. Physics for Neuromorphic Computing. *Nat. Rev. Phys.* **2020**, *2*, 499–510. [CrossRef]
2. Chua, L. Memristor-The Missing Circuit Element. *IEEE Trans. Circuit Theory* **1971**, *18*, 507–519. [CrossRef]
3. Ambrogio, S.; Balatti, S.; Nardi, F.; Facchinetti, S.; Ielmini, D. Spike-Timing Dependent Plasticity in a Transistor-Selected Resistive Switching Memory. *Nanotechnology* **2013**, *24*, 384012. [CrossRef] [PubMed]
4. Jo, S.H.; Chang, T.; Ebong, I.; Bhadviya, B.B.; Mazumder, P.; Lu, W. Nanoscale Memristor Device as Synapse in Neuromorphic Systems. *Nano. Lett.* **2010**, *10*, 1297–1301. [CrossRef] [PubMed]
5. Chanthbouala, A.; Garcia, V.; Cherifi, R.O.; Bouzehouane, K.; Fusil, S.; Moya, X.; Xavier, S.; Yamada, H.; Deranlot, C.; Mathur, N.D.; et al. A Ferroelectric Memristor. *Nature Mater.* **2012**, *11*, 860–864. [CrossRef]
6. Ambrogio, S.; Ciocchini, N.; Laudato, M.; Milo, V.; Pirovano, A.; Fantini, P.; Ielmini, D. Unsupervised Learning by Spike Timing Dependent Plasticity in Phase Change Memory (PCM) Synapses. *Front. Neurosci.* **2016**, *10*, 56. [CrossRef]
7. Kuzum, D.; Yu, S.; Wong, H.-S.P. Synaptic Electronics: Materials, Devices and Applications. *Nanotechnology* **2013**, *24*, 382001. [CrossRef]
8. Intel® Optane™ Memory H10 with Solid State Storage. Available online: https://www.intel.com/content/www/us/en/products/details/memory-storage/optane-memory/optane-memory-h10-solid-state-storage.html (accessed on 16 March 2022).
9. Wuttig, M.; Yamada, N. Phase-Change Materials for Rewriteable Data Storage. *Nature Mater.* **2007**, *6*, 824–832. [CrossRef]
10. Ji, Y.; Zeigler, D.F.; Lee, D.S.; Choi, H.; Jen, A.K.-Y.; Ko, H.C.; Kim, T.-W. Flexible and Twistable Non-Volatile Memory Cell Array with All-Organic One Diode–One Resistor Architecture. *Nat. Commun.* **2013**, *4*, 2707. [CrossRef]
11. Salvatore, G.A.; Sülzle, J.; Dalla Valle, F.; Cantarella, G.; Robotti, F.; Jokic, P.; Knobelspies, S.; Daus, A.; Büthe, L.; Petti, L.; et al. Biodegradable and Highly Deformable Temperature Sensors for the Internet of Things. *Adv. Funct. Mater.* **2017**, *27*, 1702390. [CrossRef]
12. Online Materials Information Resource—MatWeb. Available online: https://www.matweb.com/ (accessed on 1 April 2022).
13. Flexible Printed Circuit Board, Introduction and Importance. Available online: https://www.rs-online.com/designspark/flexible-printed-circuit-board-introduction-and-importance (accessed on 1 April 2022).
14. Ni, H.; Liu, J.; Wang, Z.; Yang, S. A Review on Colorless and Optically Transparent Polyimide Films: Chemistry, Process and Engineering Applications. *J. Ind. Eng. Chem.* **2015**, *28*, 16–27. [CrossRef]
15. Herth, E.; Guerchouche, K.; Rousseau, L.; Calvet, L.E.; Loyez, C. A Biocompatible and Flexible Polyimide for Wireless Sensors. *Microsyst. Technol.* **2017**, *23*, 5921–5929. [CrossRef]
16. Díaz Fattorini, A.; Chèze, C.; López García, I.; Petrucci, C.; Bertelli, M.; Righi Riva, F.; Prili, S.; Privitera, S.M.S.; Buscema, M.; Sciuto, A.; et al. Growth, Electronic and Electrical Characterization of Ge-Rich Ge–Sb–Te Alloy. *Nanomaterials* **2022**, *12*, 1340. [CrossRef] [PubMed]
17. Chèze, C.; Righi Riva, F.; Di Bella, G.; Placidi, E.; Prili, S.; Bertelli, M.; Diaz Fattorini, A.; Longo, M.; Calarco, R.; Bernasconi, M.; et al. Interface Formation during the Growth of Phase Change Material Heterostructures Based on Ge-Rich Ge-Sb-Te Alloys. *Nanomaterials* **2022**, *12*, 1007. [CrossRef] [PubMed]
18. Kumar, A.; Cecchini, R.; Wiemer, C.; Mussi, V.; De Simone, S.; Calarco, R.; Scuderi, M.; Nicotra, G.; Longo, M. MOCVD Growth of GeTe/Sb_2Te_3 Core–Shell Nanowires. *Coatings* **2021**, *11*, 718. [CrossRef]

19. Li, M.; Xie, M.; Ji, H.; Zhou, J.; Jiang, K.; Shang, L.; Li, Y.; Hu, Z.; Chu, J. PLD-Derived $Ge_2Sb_2Te_5$ Phase-Change Films with Extreme Bending Stability for Flexible Device Applications. *Appl. Phys. Lett.* **2020**, *116*, 162102. [CrossRef]
20. Zhai, F.; Liu, S.; Wang, D.; Liu, N.; Ren, Y.; Hao, Y.; Yang, K. Laser-Induced Phase Transition Processes of Amorphous Ge_2Sb_2Te5 Films. *Optik* **2019**, *185*, 126–131. [CrossRef]
21. Mun, B.H.; You, B.K.; Yang, S.R.; Yoo, H.G.; Kim, J.M.; Park, W.I.; Yin, Y.; Byun, M.; Jung, Y.S.; Lee, K.J. Flexible One Diode-One Phase Change Memory Array Enabled by Block Copolymer Self-Assembly. *ACS Nano.* **2015**, *9*, 4120–4128. [CrossRef]
22. Daus, A.; Han, S.; Knobelspies, S.; Cantarella, G.; Tröster, G. $Ge_2Sb_2Te_5$ P-Type Thin-Film Transistors on Flexible Plastic Foil. *Materials* **2018**, *11*, 1672. [CrossRef]
23. Schlich, F.F.; Wyss, A.; Galinski, H.; Spolenak, R. Cohesive and Adhesive Properties of Ultrathin Amorphous and Crystalline $Ge_2Sb_2Te_5$ Films on Polyimide Substrates. *Acta Mater.* **2017**, *126*, 264–271. [CrossRef]
24. Lu, H.; Thelander, E.; Gerlach, J.W.; Hirsch, D.; Decker, U.; Rauschenbach, B. $Ge_2Sb_2Te_5$ Phase-Change Films on Polyimide Substrates by Pulsed Laser Deposition. *Appl. Phys. Lett.* **2012**, *101*, 031905. [CrossRef]
25. Pitchappa, P.; Kumar, A.; Prakash, S.; Jani, H.; Medwal, R.; Mishra, M.; Rawat, R.S.; Venkatesan, T.; Wang, N.; Singh, R. Volatile Ultrafast Switching at Multilevel Nonvolatile States of Phase Change Material for Active Flexible Terahertz Metadevices. *Adv. Funct. Mater.* **2021**, *31*, 2100200. [CrossRef]
26. Khan, A.I.; Daus, A.; Islam, R.; Neilson, K.M.; Lee, H.R.; Wong, H.-S.P.; Pop, E. Ultralow–Switching Current Density Multilevel Phase-Change Memory on a Flexible Substrate. *Science* **2021**, *373*, 1243–1247. [CrossRef] [PubMed]
27. DigiZeitschriften: Seitenansicht. Available online: http://www.digizeitschriften.de/dms/img/?PID=GDZPPN002505045 (accessed on 4 April 2022).
28. Yamada, N.; Matsunaga, T. Structure of Laser-Crystallized $Ge_2Sb_{2+}xTe_5$ Sputtered Thin Films for Use in Optical Memory. *J. Appl. Phys.* **2000**, *88*, 7020–7028. [CrossRef]
29. Momma, K.; Izumi, F. VESTA: A Three-Dimensional Visualization System for Electronic and Structural Analysis. *J. Appl. Crystallogr.* **2008**, *41*, 653–658. [CrossRef]
30. Kozyukhin, S.A.; Nikolaev, I.I.; Lazarenko, P.I.; Valkovskiy, G.A.; Konovalov, O.; Kolobov, A.V.; Grigoryeva, N.A. Direct Observation of Amorphous to Crystalline Phase Transitions in Ge–Sb–Te Thin Films by Grazing Incidence X-Ray Diffraction Method. *J. Mater. Sci. Mater. Electron.* **2020**, *31*, 10196–10206. [CrossRef]
31. Materials Project. Available online: https://materialsproject.org/ (accessed on 30 March 2022).
32. Liu, J.; Ni, H.; Wang, Z.; Yang, S.; Zhou, W. Colorless and Transparent High—Temperature-Resistant Polymer Optical Films—Current Status and Potential Applications in Optoelectronic Fabrications. In *Optoelectronics—Materials and Devices*; Pyshkin, S.L., Ballato, J., Eds.; InTech: London, UK, 2015; ISBN 978-953-51-2174-9.
33. Nemec, P.; Nazabal, V.; Moréac, A.; Gutwirth, J.; Beneš, L.; Frumar, M. Amorphous and Crystallized Ge-Sb-Te Thin Films Deposited by Pulsed Laser: Local Structure Using Raman Scattering Spectroscopy. *Mater. Chem. Phys.* **2012**, *136*, 935–941. [CrossRef]
34. Bragaglia, V.; Holldack, K.; Boschker, J.E.; Arciprete, F.; Zallo, E.; Flissikowski, T.; Calarco, R. Far-Infrared and Raman Spectroscopy Investigation of Phonon Modes in Amorphous and Crystalline Epitaxial GeTe-Sb_2Te_3 Alloys. *Sci. Rep.* **2016**, *6*, 28560. [CrossRef]
35. Cecchi, S.; Dragoni, D.; Kriegner, D.; Tisbi, E.; Zallo, E.; Arciprete, F.; Holý, V.; Bernasconi, M.; Calarco, R. Interplay between Structural and Thermoelectric Properties in Epitaxial $Sb_{2+x}Te_3$ Alloys. *Adv. Funct. Mater.* **2019**, *29*, 1805184. [CrossRef]
36. Kapton®HN General-Purpose Polyimide Film. Available online: https://www.dupont.com/products/kapton-hn.html (accessed on 30 March 2022).
37. Zallo, E.; Wang, R.; Bragaglia, V.; Calarco, R. Laser Induced Structural Transformation in Chalcogenide Based Superlattices. *Appl. Phys. Lett.* **2016**, *108*, 221904. [CrossRef]
38. Kato, T.; Tanaka, K. Electronic Properties of Amorphous and Crystalline $Ge_2Sb_2Te_5$ Films. *Jpn. J. Appl. Phys.* **2005**, *44*, 7340–7344. [CrossRef]
39. Lazarenko, P.I.; Sherchenkov, A.A.; Kozyukhin, S.A.; Babich, A.V.; Timoshenkov, S.P.; Gromov, D.G.; Shuliatyev, A.S.; Redichev, E.N. Electrical Properties of the Ge2Sb2Te5 Thin Films for Phase Change Memory Application. *AIP Conf. Proc.* **2016**, *1727*, 020013. [CrossRef]
40. Yakubov, A.; Sherchenkov, A.; Lazarenko, P.; Babich, A.; Terekhov, D.; Dedkova, A. Contact resistance measurements for the $Ge_2Sb_2Te_5$ thin films. *Chalcogenide Lett.* **2020**, *17*, 1–8.
41. Raoux, S.; Rettner, C.T.; Jordan-Sweet, J.L.; Kellock, A.J.; Topuria, T.; Rice, P.M.; Miller, D.C. Direct Observation of Amorphous to Crystalline Phase Transitions in Nanoparticle Arrays of Phase Change Materials. *J. Appl. Phys.* **2007**, *102*, 094305. [CrossRef]
42. Guo, P.; Sarangan, A.; Agha, I. A Review of Germanium-Antimony-Telluride Phase Change Materials for Non-Volatile Memories and Optical Modulators. *Appl. Sci.* **2019**, *9*, 530. [CrossRef]
43. Bragaglia, V.; Arciprete, F.; Zhang, W.; Mio, A.M.; Zallo, E.; Perumal, K.; Giussani, A.; Cecchi, S.; Boschker, J.E.; Riechert, H.; et al. Metal—Insulator Transition Driven by Vacancy Ordering in GeSbTe Phase Change Materials. *Sci. Rep.* **2016**, *6*, 23843. [CrossRef]
44. Wong, H.-S.P.; Raoux, S.; Kim, S.; Liang, J.; Reifenberg, J.P.; Rajendran, B.; Asheghi, M.; Goodson, K.E. Phase Change Memory. *Proc. IEEE* **2010**, *98*, 2201–2227. [CrossRef]

Article

Effect of Nitrogen Doping on the Crystallization Kinetics of Ge$_2$Sb$_2$Te$_5$

Minh Anh Luong [1], Nikolay Cherkashin [1], Béatrice Pecassou [1], Chiara Sabbione [2], Frédéric Mazen [2] and Alain Claverie [1,*]

[1] CEMES-CNRS, 29 Rue Jeanne Marvig, 31055 Toulouse, France; minh-anh.luong@cemes.fr (M.A.L.); nikolay.cherkashin@cemes.fr (N.C.); beatrice.pecassou@cemes.fr (B.P.)

[2] Léti-CEA, 17 Avenue des Martyrs, F-38000 Grenoble, France; chiara.sabbione@cea.fr (C.S.); frederic.mazen@cea.fr (F.M.)

* Correspondence: alain.claverie@cemes.fr

Abstract: Among the phase change materials, Ge$_2$Sb$_2$Te$_5$ (GST-225) is the most studied and is already integrated into many devices. N doping is known to significantly improve some key characteristics such as the thermal stability of materials and the resistance drift of devices. However, the origin, at the atomic scale, of these alterations is rather elusive. The most important issue is to understand how N doping affects the crystallization characteristics, mechanisms and kinetics, of GST-225. Here, we report the results of a combination of in situ and ex situ transmission electron microscopy (TEM) investigations carried out on specifically designed samples to evidence the influence of N concentration on the crystallization kinetics and resulting morphology of the alloy. Beyond the known shift of the crystallization temperature and the observation of smaller grains, we show that N renders the crystallization process more "nucleation dominated" and ascribe this characteristic to the increased viscosity of the amorphous state. This increased viscosity is linked to the mechanical rigidity and the reduced diffusivity resulting from the formation of Ge–N bonds in the amorphous phase. During thermal annealing, N hampers the coalescence of the crystalline grains and the cubic to hexagonal transition. Making use of AbStrain, a recently invented TEM-based technique, we evidence that the nanocrystals formed from the crystallization of N-doped amorphous GST-225 are under tension, which suggests that N is inserted in the lattice and explains why it is not found at grain boundaries. Globally, all these results demonstrate that the origin of the effect of N on the crystallization of GST-225 is not attributed to the formation of a secondary phase such as a nitride, but to the ability of N to bind to Ge in the amorphous and crystalline phases and to unbind and rebind with Ge along the diffusion path of this atomic species during annealing.

Keywords: phase change materials; Ge$_2$Sb$_2$Te$_5$; nitrogen; crystallization; strain; kinetics

1. Introduction

Phase change materials (PCMs) are materials which show dramatic variations of several of their physical properties, such as the optical reflectance and electrical resistivity, which result from a change in their structure from amorphous to the crystalline states [1–5]. After their successful exploitation in compact disc read-only memories (CD ROMs), PCMs are currently exploited in phase change random-access memories (PC-RAMs), where the bit of information is encoded within two distinct resistive states corresponding to the (high-resistive) amorphous state and the (low-resistive) crystalline state. Reversible SET to RESET transitions are obtained by feeding a cell with appropriate pulses of electrical current and heating the material, generally a small dome of a few tens of nanometers in diameter, to crystallize it or, alternatively, to quench it from the melt.

The high resistivity contrast and the fast switching between these states offered by Ge$_2$Sb$_2$Te$_5$ (GST-225) have motivated its integration into high-performance digital devices [6,7]. Moreover, recent reports have demonstrated the possibility to program multilevel cells using

GST-225, a promising step towards their integration as synaptic elements in artificial neural networks, as needed for neuromorphic computing [8–10].

However, GST-225 shows several characteristics which severely limits the application field of devices using it. Of first concern is its limited thermal stability. Its crystallization temperature (from 120 to 180 °C, depending on purity, homogeneity and layers in contact) is too low to preserve code integrity during soldering processes, as needed for embedded applications, and to ensure good data retention under moderate temperature conditions. Another drawback is the tendency of the resistivity of the RESET state to drift over time, probably the result of some structural relaxation of the amorphous phase obtained after quenching from the melt [11]. For these reasons, there is an increasing demand for PCMs exhibiting better thermal stability.

Doping with chemical impurities may bring solutions. Carbon [12], oxygen [13], bismuth [14] and antimony [15] have been reported to increase the crystallization temperature of GST-225. Doping with a few percent of nitrogen is also appealing because it represents an effective way to achieve much higher crystallization temperatures (T_x) [16–19], increased resistivity of both the crystalline and amorphous states while maintaining a high contrast [20–24], and a reduced resistivity drift of the RESET state [25,26]. Moreover, N has also been shown to render the transition from the amorphous to the crystalline states more progressive, giving more precise access to intermediate resistivity states between the RESET and SET values [21].

There are many reports on the effect of N on the thermal crystallization and resulting microstructure of GST-225. There is an overall consensus that T_x gradually increases with N concentration and that a maximum crystallization temperature of about 250 °C can be reached for concentrations in the 8–12% range [16,20–24,27–31]. However, for much higher concentrations, the Ge-poor $Ge_1Sb_2Te_4$ phase crystallizes instead of the desired GST-225 [16]. Moreover, N doping also seems to shift the cubic to hexagonal transition to a much higher temperature, classically occurring at about 300 °C in pure GST-225 [16,29]. Another noticeable effect of N doping is a significant reduction in the size of the grains which are formed after crystallization [22,24,28,31].

Again, although these effects have been well established, their origin is unclear. As the thermal crystallization of GST-225 proceeds through different steps, incubation, nucleation and growth, one reason for this limited understanding lies in the difficulty to infer the mechanisms which are impacted by N during crystallization based on the sole observation of the microstructure of the material after annealing. Alternatively, Privitera et al. [23], while using resistivity measurements, a quite indirect technique, have shown that in situ measurements during annealing provide more insights into those mechanisms, revealing how their respective kinetics are impacted. Thus, we believe that the observation and recording of the full sequence of crystallization of GST-225 during in situ annealing in a transmission electron microscope (TEM) would be very much appropriate to pinpoint the impact of N on the nucleation and growth of the crystalline phase.

Another issue is that, most often and whatever the characterization technique, the influence of N is studied through the comparison of the results obtained on a limited number of layers doped at different N concentrations. If in situ annealing in the TEM is to be used, samples of different concentrations are to be annealed in situ one by one, rendering the extraction of reliable data for comparison, at best, delicate. To circumvent this difficulty, samples of doped and undoped GST-225 should be preferably annealed at the same time, during the same in situ annealing experiment. This can be achieved by preparing specifically designed layers by combining deposition and ion implantation techniques. Through the careful selection of layer thickness, ion beam energy and fluence, a reasonably thick layer consisting of a N-doped region on top a pristine amorphous GST-225 can be fabricated. Moreover, the concentration profile resulting from the ion implantation may be exploited to assess the influence of the concentration over a large range of values.

Given the remarks above, we have decided to combine the advantages provided by in situ annealing experiments in the TEM and these N-implanted GST-225 layers. This work reports the direct imaging of the crystallization sequences which affect amorphous GST-225 during annealing and evidences the morphological and kinetic differences due to N doping. This allows for a fact-based discussion on the underlying mechanisms.

The second point to clarify the atomic location of N. There is experimental evidence that, in the amorphous phase, N preferentially binds to Ge, eventually forming nitrides for high N concentrations [21,22,32]. In the crystalline phase, there is no consensus. Simulations tend to suggest that inserting N into crystalline GST-225 is too costly and that it will be expelled at crystallization [33]. This reinforced the widespread belief that N, in the form of nitrides, resides at the grain boundaries then inhibits the further growth of these grains [28,33–35]. However, one has to note that, although nitride-like characteristics have been readily evidenced in crystalline N-doped GST-225 [23,32], there are no reports showing TEM images or other direct experimental evidence of such phases decorating the GST-225 grain boundaries. On the contrary, there are several reports showing, by X-ray diffraction (XRD), the increase in the lattice parameters of polycrystalline GST-225 layers for increasing N concentrations. The interpretation of data differs, however, as well as the range of concentrations investigated, assigning this characteristic either to the occupation of vacancy sites [24] in the (Ge, Sb, V) sub-lattice or to the insertion of N into tetrahedral sites [28,30].

It is to be noted that all these results were obtained from a very limited number of samples, in terms of annealing conditions and N concentrations. They were also obtained on fully crystallized layers. Ideally, measurements aimed at evidencing the effect of N incorporation on the strain state of crystalline GST-225 should be carried out on single grains, before they come in contact and possibly exert stress on each other, i.e., in samples where the material is only partially crystallized.

For this reason, we have used a recently invented TEM-based technique, named "AbStrain", initially developed to correct experimental high resolution (HR) TEM images from distortions and calibration errors and for mapping the exact interplanar distances and angles in crystals with high precision [36]. Here, this technique is used to measure the changes of interplanar spacing (strain) of small GST-225 nanocrystals resulting from the eventual incorporation of N upon crystallization.

Thanks to a unique combination of advanced techniques of TEM and specifically designed samples, in this paper we show that the incorporation of N renders the crystallization of GST-225 more dominated by nucleation, a characteristic to be ascribed, along with the much larger crystallization temperature, to the increased viscosity of the N-doped amorphous GST. Moreover, we evidence that small grains crystallizing in N-doped GST-225 show strong positive strain, hence suggesting the direct incorporation of N into the GST lattice.

2. Experimental

A 500 nm thick $Ge_2Sb_2Te_5$ film was grown in the amorphous state by physical vapor deposition on a naturally oxidized 300 mm silicon (100) wafer using an industrial tool. The $Ge_2Sb_2Te_5$ film was capped in situ with a 20 nm thick GeN layer to protect it from oxidation. The wafer was cut into pieces then implanted with nitrogen ions at 80 keV with a fluence of 3.8×10^{16} ions/cm^2. These conditions were intended to incorporate a maximum concentration of about 5% (atomic fraction) at a depth of about 150 nm from the surface and a concentration decreasing down to almost 0% at the surface and at a depth of about 300 nm. A 200 nm thick bottom part of the 500 nm thick GST layer was thus left undoped, providing a reference in the same sample. Preliminary characterization of the as-implanted films has evidenced that ion implantation may cause the recrystallization of the layer, due to heating and collisional effects. For this reason, a very low beam current of 40 µA was used for implantation, for which the layer was checked to have remained fully amorphous (Supplementary Materials Figure S2). Subsequently, 1×1 cm^2 specimens were annealed in a horizontal Carbolite furnace under atmospheric pressure and N_2 gas flow for temperatures ranging from 170 to 300 °C and times from 30 min to 1 h. Thin samples of the annealed layers, suitable for cross-sectional transmission electron microscopy (XTEM) observations, were prepared by focus ion beam (FIB) using an FEI Helios NanoLab 600 (FEI Company) operating with a 30 keV Ga ion beam and finally polished and cleaned at 2 keV. The samples were imaged and analyzed by various TEM techniques using either an aberration-corrected FEI TECNAI F20 (200 KeV), a Philips CM20-FEG (200 KeV) or the I^{2}TEM from Hitachi (300 KeV) [37].

In parallel, the thermal crystallization of the implanted and non-implanted films was observed during in situ annealing in a TEM, in bright-field and dark-field modes, and recorded. This in situ heating was performed using a Gatan 652 double tilt heating holder which was connected to a temperature controller and acted as a furnace-type holder. In the experiments presented here, the temperature was increased by steps of 5 °C from about 100 °C, using a ramping rate of 1 °C/s, and held for 5 min at this temperature during which the film was imaged (Figure 1). Special attention has been paid to reduce the influence of electron beam irradiation on the crystallization by minimizing the exposure time and beam intensity [29]. Moreover, fresh Ge$_2$Sb$_2$Te$_5$ areas were also analyzed and compared to those left under the beam.

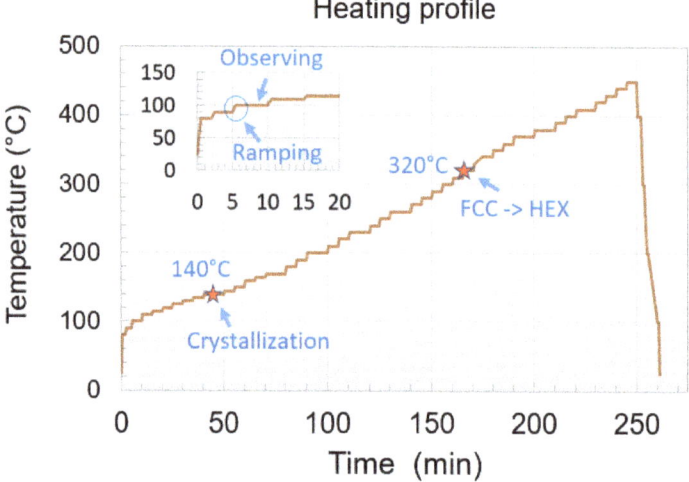

Figure 1. Typical heating profile used for in situ TEM annealing. The temperature was increased by steps of 5 °C from about 100 °C, using a ramping rate of 1 °C/s, and held for 5 min at this temperature during which the film was imaged (shown by the insertion). FCC to HEX refers to the transition from the face-centered cubic to the hexagonal phase.

3. Results and Discussion

3.1. Crystallization Kinetics

3.1.1. Pristine Sample

In situ TEM is a powerful technique, but artefacts may affect the kinetics and even the type of the observed phenomena, due to the limited specimen thickness, its possible oxidation, electron beam irradiation and heating effects. For this reason, before giving credit to the results obtained when studying the thermal behavior of N-implanted Ge$_2$Sb$_2$Te$_5$ in situ, we first investigated the crystallization of pure and amorphous Ge$_2$Sb$_2$Te$_5$ for referencing. The structure and stoichiometry of the as-deposited Ge$_2$Sb$_2$Te$_5$ were checked by TEM and energy dispersive X-ray spectroscopy (EDX), and the results are shown in the Supplementary Materials Figure S1.

Figure 2 is a montage of snapshots summarizing the main results which were extracted from a video taken during the heating of the specimen. Up to 135 °C, nothing happened. At 140 °C, small crystals started to nucleate homogeneously within the layer (indicated by white arrows). They were small, typically from 2 to 8 nm in diameter (mean diameter of 5 nm). With increasing time and temperature, these grains grew while more grains nucleated.

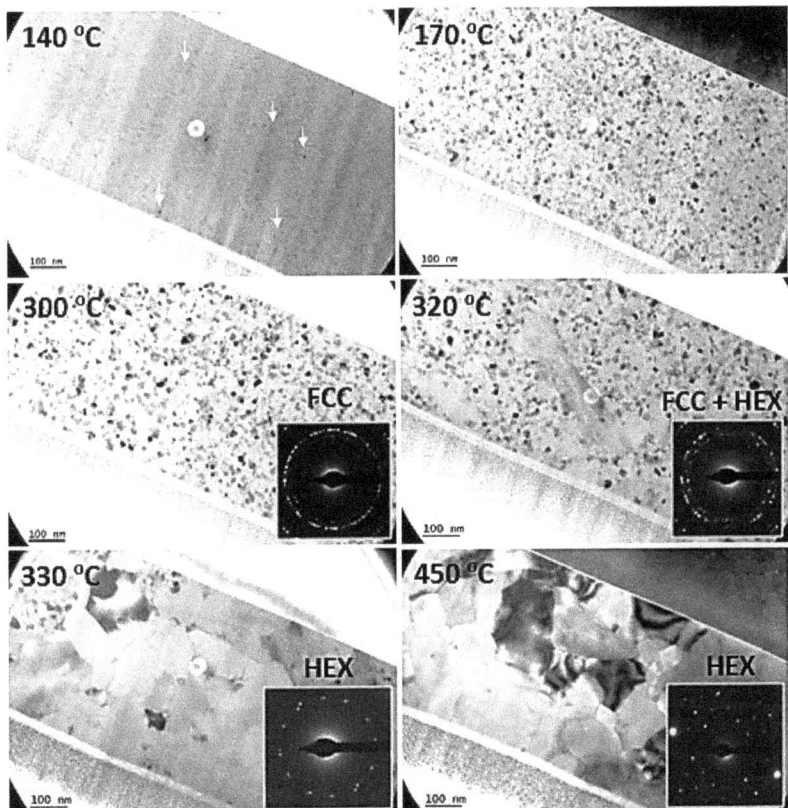

Figure 2. Bright-field (BF) TEM micrographs showing the structure of a pristine $Ge_2Sb_2Te_5$ layer during the increase in the annealing temperature during the in situ experiment. The inserts show the associated SAED patterns. "FCC" and "HEX" refer to the face-centered cubic and hexagonal structures, respectively.

From about 170 °C, i.e., after having spent about 30 min above 140 °C, the layer was, for a large part, crystalline, as deduced from the selected area electron diffraction (SAED) pattern and the diffraction contrast in the image. Beyond this temperature, the analysis of the grain population revealed that the grains continuously increased in size during ramping up (see Table 1), up to 300 °C. Electron diffraction patterns show the expected face-centered cubic (FCC) structure signature of the GST-225 grains. This regular and quite slow growth regime is thought to mostly result from the crystallization of the remaining amorphous material, although the competitive growth of existing nanocrystals may also contribute. However, from 300 °C and above, coalescence of the grains was observed, and consequently, the size of the grains dramatically increased. It is interesting to note that the new large grains which resulted from this coalescence showed the hexagonal stable structure of GST-225. From 330 °C and above, all grains were quite large (a few hundred nanometers) and showed a hexagonal structure. They did not evolve significantly in size but showed better defined grain boundaries when heated up to 450 °C (see Supplementary Materials Video S1–S4 for the real-time observation of the evolution of the structure).

Table 1. Mean Ge$_2$Sb$_2$Te$_5$ grain sizes as functions of annealing temperature.

Temperature (°C)	140	160	170	180	200	230	270	300
Grain Size (±3 nm)	5	10	12	15	18	20	23	30

Thus, the information that we could extract from our in situ TEM observation of the crystallization of stoichiometric GST-225 was in very good agreement with those reported in the literature [2–4]. One could probably argue that the crystallization temperature that we observed, 10 to 20 °C below the most commonly reported values, may be due to the oxidation of the FIB specimen or to electron beam irradiation. Oxidation results in the heterogeneous and fast nucleation of the crystalline phase from the oxidized layer [38], a phenomenon which we have observed on intentionally oxidized layers but not in the experiments reported here. Electron irradiation may also facilitate the nucleation and growth of GST grains [29]. However, we have systematically compared images obtained from "fresh" and beam-exposed areas without detecting quantifiable differences, thanks to the experimental precautions under which the specimens were imaged. Instead, we prefer to stress that, usually, the crystallization temperature is identified through resistivity measurements during fast and continuous heating of the material, typically 10 °C/min. In our experiments, the heating rate was high (1 °C/s), but the sample was maintained at constant temperature for 5 min after every 5 °C jump. Thus, on average, more time was left for the material to incubate and nucleate the crystalline phase, which we think is the reason for it presenting a slightly lower crystallization temperature.

However, this test-study demonstrates that, provided basic precautions are respected, the in situ heating of thin GST lamellas in a TEM is a reliable technique to investigate the details of the crystallization phenomenon.

3.1.2. N-Implanted Samples

Figure 3 is a montage of snapshots showing the behavior and structural characteristics of the N-implanted layer during the same type of in situ annealing. From 140 °C, the nucleation of the crystalline phase is observed to start, as in the pristine sample, but only in the very surface region and at depths larger than 300 nm, i.e., in the regions where the N concentration is low, but not in the implanted region. When increasing the temperature up to 250 °C, these GST crystalline grains grew while the implanted region from below the surface towards a depth of approximately 300 nm remained amorphous. It was only when reaching 250 °C that the first crystalline nuclei started to appear in this region. At 330 °C, grains in the unimplanted bottom region dramatically grew and progressively transformed into the hexagonal phase. They even started growing from their initial location into the implanted region of the layer. Interestingly, the grains located in the doped region appeared to grow more slowly that those located in the bottom part. Finally, when reaching 400 °C, the layer was fully crystalline but still showed clear grain size differences depending on whether the grains sat in the N-doped or undoped regions.

To clarify the impact of N on the growth of the GST grains, we have measured their sizes in samples annealed ex situ under well-controlled conditions, which is more adapted and precise than from fast-evolving BF images taken during in situ annealing. Figure 4 shows, as a typical example, a set of BF and dark-field (DF) XTEM images obtained on the sample annealed at 180 °C for 30 min. The DF image clearly reveals that the grain sizes are distributed along the depth, in correspondence to the concentration of implanted nitrogen, as shown by the implanted profile (Figure 4c) extracted from the Monte Carlo simulation of the implantation [39].

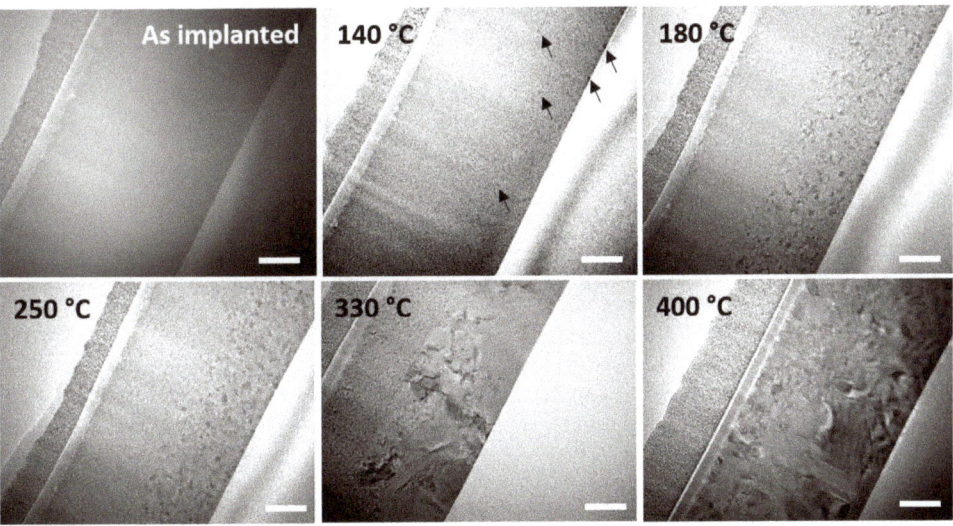

Figure 3. BF TEM micrographs showing the evolution of the structure of the N-implanted GST-225 layer during in situ annealing. The white scale bars refer to a length of 100 nm.

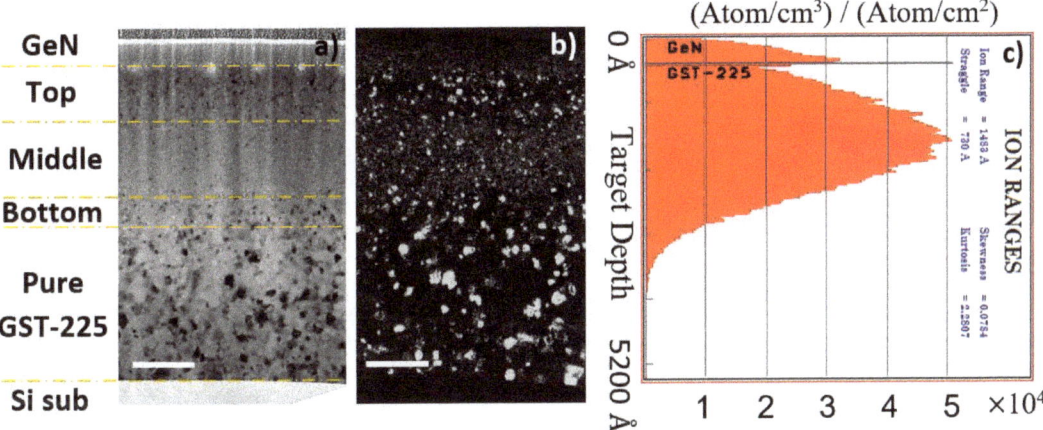

Figure 4. BF (**a**) and dark-field (DF) (**b**) images of the N-implanted GST-225 layer, annealed at 180 °C for 30 min. The N-implanted region is divided into different regions, the top, middle and bottom which are N-implanted, and the unimplanted region which provides the pristine reference. (**c**) SRIM simulation of the nitrogen depth-distribution after implantation. The white scale bars refer to a length of 100 nm.

To analyze such images, we divided the specimen into four regions: the top, extending from the GeN/GST interface to a depth of approximately 80 nm; the middle, from this depth to about 280 nm, i.e., centered on the projected range of the ions (Rp); the bottom, from this depth to about 320 nm; and the unimplanted region, which extended towards the substrate.

Table 2 shows the results of the statistical analysis of the grain sizes in each of these regions, as a function of annealing conditions.

Table 2. Mean GST-225 grain sizes as a function of depth position in the layer and for different annealing temperatures. The BF and DF images used for data extraction are shown in the Supplementary Materials Figure S3.

	Top	Middle (Rp Region)	Bottom	Unimplanted
175 °C—30 min	0 (few isolated)	0 (none)	0 (few isolated)	18 nm ± 2 nm
180 °C—30 min	6 nm ± 2 nm	2 nm ± 2 nm	6 nm ± 2 nm	20 nm ± 2 nm
210 °C—30 min	8 nm ± 2 nm	6 nm ± 2 nm	8 nm ± 2 nm	20 nm ± 2 nm
250 °C—30 min	10 nm ± 2 nm	10 nm ± 2 nm	10 nm ± 2 nm	22 nm ± 2 nm
300 °C—1 h	10 nm ± 2 nm	10 nm ± 2 nm	10 nm ± 2 nm	>80 nm

In the unimplanted region, the grains had a mean size of about 18–22 nm after annealing at 170 °C for 30 min and did not grow further when increasing the annealing temperature until they coalesced at 300 °C and above, showing a sudden and dramatic increase in the grain size. In contrast, in the middle region, where the N concentration was above 2%, grain nucleation was not activated after 170 °C/30 min annealing and only started after annealing at 180 °C. Increasing the annealing temperature allowed for the growth of these grains only up to a size of about 10 nm, and did not evolve further even when annealing at 300 °C for 1 h. In the top and bottom regions, where the N concentration ranged from 1% to 2%, the situation was intermediate; the nucleation only started after 175 °C/30 min annealing and the grains further grew when increasing the temperature, but only up to a size of about 10 nm after 300 °C/1 h annealing.

4. Discussion

The experimental results reported above show that the doping by N of amorphous GST-225 does not change its crystallization mechanism. Crystallization still proceeds through the homogeneous nucleation of grains followed by their growth. However, we have evidenced that, in the presence of N, the nucleation regime requires a higher temperature to be activated. Most importantly, we observe that, at the end of the crystallization process, when the whole volume is totally crystallized, the grains are smaller and more numerous in the N-doped region (see Table 2, 300 °C, 1 h). This characteristic cannot result from heterogeneous nucleation on N-related sites, which would have led to a decrease in the nucleation temperature; therefore, it must be ascribed to the fact that, in the temperature range studied here, the nucleation rate is high while the growth rate is comparatively low. The higher the N concentration, the more pronounced is this effect.

In Figure 5, we have schematically compared the crystallization characteristics of undoped and N-doped GST-225. From stress experiments, we know that T_g, the glass transition temperature, increases with N doping [27]. In contrast, there is no evidence that T_m, the melting temperature of GST-225, is impacted. There are even reports showing that the Tm of GeTe is unaffected by N doping [40]. We have evidenced that T_x, the crystallization temperature, is dramatically increased in N-doped GST-225. Thus, T_g and T_x are shifted in the diagram related to N-doped GST. Actually, from our observations, beyond the shift towards higher temperatures of both the nucleation and growth probabilities, the main effect of N is to increase the ratio between the nucleation and the growth probabilities. This figure illustrates that, in N-doped GST, the crystallization mechanism is more "nucleation dominated" than in pure GST, in the range of temperatures investigated in this work.

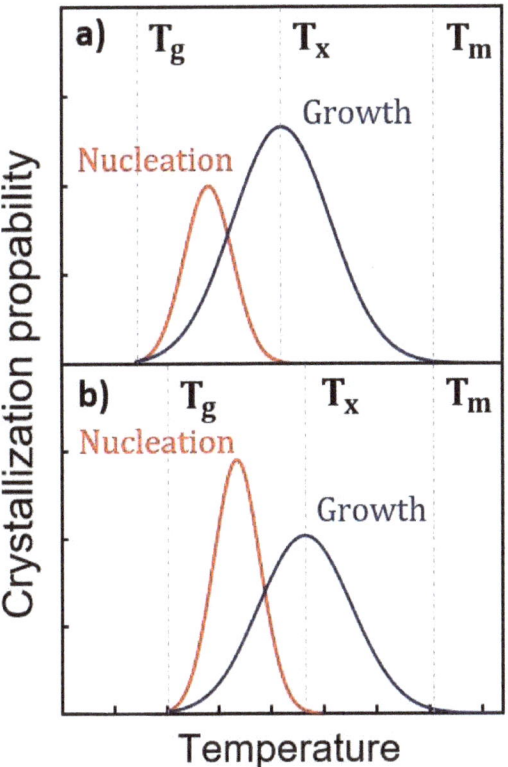

Figure 5. Schematic illustrations of nucleation dominated kinetics for undoped GST-225 (**a**) and N-implanted GST-225 (**b**). T_g, T_x and T_m are the glass transition, the crystallization and the melting temperatures, respectively.

From the classical nucleation theory, we understand that this characteristic results from the increase in the viscosity of the amorphous material when doped by nitrogen [3,41–43]. This viscosity increase must be related to the mechanically constrained environment of the Ge atoms, some of which are bonded to N [44]. The viscosity of a glass is inversely proportional to the diffusivity of the atomic species that compose it; our results can also be interpreted as being due to the reduction in Ge diffusivity in the presence of N, in the amorphous phase.

Moreover, the observed resistance of the grain to coalescence should be mentioned. The coalescence of neighboring grains requires the collaborative motion of several tens of thousands of atoms. Again, we can ascribe this resistance to the reduced diffusivity of at least one of the constituents of the material, most probably Ge. However, this occurs in the crystalline phase and, up to now, the exact location of N has remained elusive.

Strain in the Grains: N Location

A second important question concerns the location of N in the GST material. It is now clear that, in the amorphous phase, N is most probably bound to Ge, which explains the increased viscosity of the material. However, in the crystalline state, the question remains open.

There have been a number of reports evidencing, most often by X-ray photoelectron spectroscopy (XPS), the presence of a Ge nitride phase after the crystallization process [23,32]. The formation of such a phase, sitting preferentially at the grain boundaries

(GBs), could eventually explain the reduction in the growth rate. However, if the 5% of nitrogen initially contained in the amorphous material has transformed into Ge_3N_4, this phase should be detected. Moreover, the stoichiometric imbalance created by the formation of a second phase involving Ge and N should lead to the formation of Ge-poor GST phases, such as the $Ge_1Sb_2Te_4$.

Actually, despite our efforts and similarly to Song et al. [45], we have not been able to detect such a phase by EDX or EELS, nor we have noticed any precipitate or thin layer decorating the GBs. We could not evidence the $Ge_1Sb_2Te_4$ phase either, and all grains show the expected GST-225 cubic or hexagonal structures. On the contrary, the chemical imaging of the crystallized layer with a resolution of about 2 nm suggests that nitrogen is homogeneously dispersed in the grains. If nitrogen, bound to Ge or as molecular N_2, resides inside the grains, these grains should show some deformation with respect to their regular interplanar spacings. To check this possibility, we have carried out strain measurements, comparing the lattice spacings and characteristic angles in the grains, depending on whether they were found in the N-doped or undoped regions. This situation, where isolated nanocrystals of random orientation are buried in an amorphous matrix and located relatively far from any reference crystal, is well beyond the application fields of all popular strain measurement techniques, working in the image [46,47] or diffraction modes [48,49]. For this reason, we have used the novel AbStrain technique specifically invented to overcome these limitations. Its working principle is described elsewhere [36,50]. We briefly recall the operational procedure which was used. To ensure that the strain that we could eventually evidence mostly resulted from N doping and not from some stress generated by the contact between neighboring grains, we focused our attention on the sample annealed at 210 °C for 30 min, in which the nanocrystals were still growing at the expense of the remaining amorphous matrix.

First, a reference HR-TEM image of Si lattice was taken along the $[110]^{Si}$ zone axis. This image was used for the measurement of the systematic image calibration errors and distortions arising from the microscope and charge-coupled device (CCD) camera. Secondly, HR-TEM images of many GST crystals were acquired using the same magnification, tilt and defocus conditions, in the unimplanted then in the N^+-implanted zones, by shifting the sample. Afterwards, the GST crystals showing the fringe contrasts expected from the structure when viewed along the $<001>^{GST}$, $<112>^{GST}$ or $<111>^{GST}$ zone axes were selected for further analysis. After correcting from systematic errors and distortions arising from the microscope, exact measurements of the lattice spacings and angles were carried out (in the Fourier space) on nanocrystals located both in the implanted and unimplanted regions. "Absolute" strain tensor components, i.e., components defined with reference to the perfect bulk GST lattice (as from ICDS file), could then be calculated [36].

Figure 6 shows an HR-TEM image of the reference Si lattice (Figure 6a), an HR-TEM image of a large $Ge_2Sb_2Te_5$ crystal located in the unimplanted region and viewed along the $<001>^{GST}$ zone axis (Figure 6b), and the result of the AbStrain analysis of this image (Figure 6c,d). Figure 6c shows the maps of the (200) and (020) interplanar distances and of the angle between these planes. From them, the maps of three strain tensor components, defined with reference to the perfect GST-225 cubic lattice, can be calculated (Figure 6d). The average measured values of d_{200} = 0.302581 nm, d_{020} = 0.301462 nm and $\angle(g_1;g_2)$ = 90.7179° are very close to the values characteristic of the relaxed GST lattice expected at d_{200} = 0.301843, d_{020} = 0.301843 and $\angle(g_1;g_2)$ = 90°. Consequently, all the three measured strain components were very close to zero. This is a clear evidence that the large nanocrystals located in the unimplanted region of the layer are made of pure and relaxed $Ge_2Sb_2Te_5$ material.

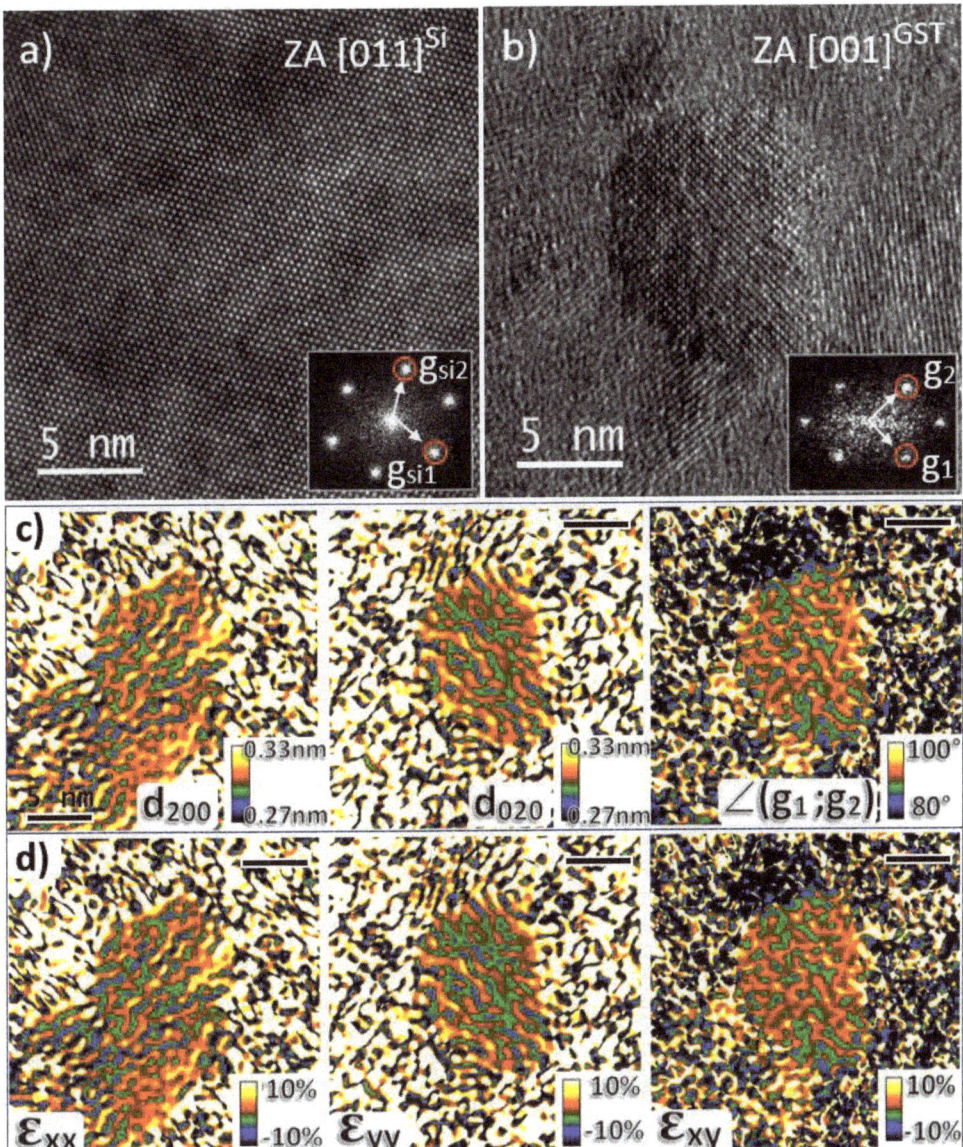

Figure 6. HR-TEM images of the reference Si lattice (**a**) and of a large Ge$_2$Sb$_2$Te$_5$ crystal viewed along the <001>^{GST} zone axis (**b**). Inserts show their corresponding FFTs and the reciprocal vectors used for AbStrain analysis. AbStrain analysis: maps of the (200) and (020) interplanar distances and of the angle between these planes obtained after correction (**c**). Extracted strain components ε_{xx}, ε_{yy} and ε_{xy} (reference is the perfect GST-225 cubic lattice) (**d**). The black scale bars refer to a length of 5 nm.

The same analysis was carried out on 15 different nanocrystals located in the same region and the results are plotted in Figure 7a. The shear strain components were always close to 0 (<0.05%) and are not shown. The plots of ε_{xx} and ε_{yy} show that their mean values are close to zero (at about −0.1%), and that all measured strain values are below +/−1%, with most of them below +/−0.5%, which must result from the limitations of the technique and from the actual strain states of the nanocrystals.

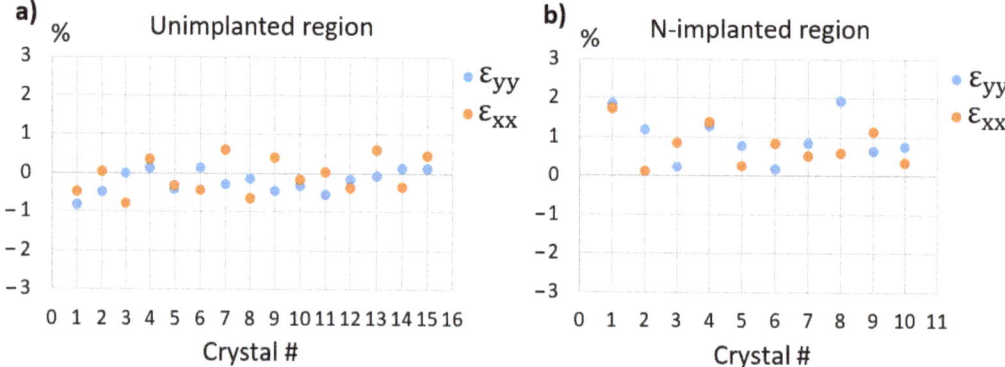

Figure 7. Strain components (ε_{xx}, ε_{yy}) extracted from 15 large crystals in the unimplanted region (**a**) and 10 smaller crystals in the N-implanted region (**b**). The specimen was annealed at 210 °C for 30 min to initiate crystallization.

Figure 7b plots the results we have obtained on 10 small nanocrystals located in the implanted region. In strong contrast to what was measured in the unimplanted region, large strain values, up to 1.8%, were measured in most of these nanocrystals. Moreover, all these values were found to be positive (mean value: 0.7%), evidence that the nanocrystals are in tension. Clearly, the most probable explanation for this overall positive strain is the presence of N inside the grains, as a bound interstitial or inserted molecule [51]. Whether these N atoms sit between FCC planes or partly fill the vacancies available on one half of the {111} planes would result in the same positive dilatation of the crystalline network and cannot be inferred from our results.

5. Conclusions

In conclusion, using specifically designed samples, we have used a combination of in situ and ex situ TEM techniques to study the influence of nitrogen on the thermal crystallization of GST-225. The samples were obtained by nitrogen ion implantation in such a way that a Gaussian-like concentration profile centered at a depth of about 150 nm with a maximum N concentration of 5% was introduced in the 500 nm thick layers. In situ annealing experiments evidenced the need for a higher temperature for nucleating the GST-225 crystalline phase when the N concentration is larger than about 1%. This effect becomes more pronounced as the N concentration increases, with a maximum shift of about 50 °C for 5% of N. Moreover, the growth rate of the grains is limited in the presence of N and crystallization occurs mostly though the nucleation of new grains. As a result, when the material is totally crystallized, grains are smaller and present in larger densities. Thus, these observations show that the effect of N doping is to render the crystallization of the material more "nucleation dominated". This can be understood as resulting from the binding of N with Ge which, inducing mechanical rigidity to the amorphous network and reducing Ge diffusivity, increases its viscosity [44]. This decrease in Ge diffusivity may also explain the observed resistance of the GST grains to coalescence and to transit from the FCC to the hexagonal phases in the presence of N, during high-temperature annealing.

The location of N in the crystalline phase was quite controversial, and our processors [45] were not able to detect any nitride phase decorating the grain boundaries; therefore, we used a recently invented technique, AbStrain, to measure the lattice spacings of nanocrystals formed by crystallizing N-rich amorphous GST. We have clearly demonstrated that all these nanocrystals are subjected to quite a large positive strain. They are under tension, which strongly suggests that N must be inserted in the lattice.

Globally, all these results demonstrate that the origin of the effect of N on the crystallization of GST-225 is not attributable to the formation of a secondary phase, eventually stable at room temperature, for example a nitride, but to the ability of N to bind to Ge in the amorphous and crystalline phases and to unbind and rebind with Ge along the diffusion path of this atomic species during annealing.

Supplementary Materials: The following are available online at https://www.mdpi.com/article/10.3390/nano11071729/s1, Video S1: In situ annealing for a pure $Ge_2Sb_2Te_5$ specimen, taken at 320 °C (25 frame/s, speed × 15), Video S2: In situ annealing for a N-implanted $Ge_2Sb_2Te_5$ specimen, taken at 140 °C (25 frame/s, speed × 15), Video S3: In situ annealing for a N-implanted $Ge_2Sb_2Te_5$ specimen, taken at 320 °C (25 frame/s, speed × 15), Video S4: In situ annealing for a N-implanted $Ge_2Sb_2Te_5$ specimen, taken at 140 °C (25 frame/s, speed × 15), Supplementary Materials, Figure S1 shows the BF image of an as-deposited 500 nm thick $Ge_2Sb_2Te_5$ specimen, and EDX mapping for the elements present, Supplementary Materials, Figure S2 shows the BF image of a N-implanted $Ge_2Sb_2Te_5$ specimen, and EDX mapping for the elements present, Supplementary Materials, Figure S3 shows BF–DF images of a N-implanted $Ge_2Sb_2Te_5$ specimen, annealed at different temperatures and durations (i.e., 175, 180, 210 and 250 for 30 min, and 300 °C for 1 h).

Author Contributions: Methodology and investigation, M.A.L. and N.C.; resources, B.P., C.S. and F.M.; conceptualization and funding acquisition, A.C. All authors have read and agreed to the published version of the manuscript.

Funding: This work is part of the "Ô-GST Project" and partially funded by the nano2022 (IPCEI) initiative.

Data Availability Statement: Data available in a publicly accessible repository.

Conflicts of Interest: The authors declare no conflict of interest.

References

1. Wuttig, M.; Yamada, N. Phase-Change Materials for Rewriteable Data Storage. *Nat. Mater.* **2007**, *6*, 824–832. [CrossRef] [PubMed]
2. Lee, B.C.; Zhou, P.; Yang, J.; Zhang, Y.; Zhao, B.; Ipek, E.; Mutlu, O.; Burger, D. Phase-Change Technology and the Future of Main Memory. *IEEE Micro* **2010**, *30*, 143. [CrossRef]
3. Raoux, S.; Wełnic, W.; Ielmini, D. Phase Change Materials and Their Application to Nonvolatile Memories. *Chem. Rev.* **2010**, *110*, 240–267. [CrossRef] [PubMed]
4. Noé, P.; Vallée, C.; Hippert, F.; Fillot, F.; Raty, J.-Y. Phase-Change Materials for Non-Volatile Memory Devices: From Technological Challenges to Materials Science Issues. *Semicond. Sci. Technol.* **2018**, *33*, 013002. [CrossRef]
5. Guo, P.; Sarangan, A.; Agha, I. A Review of Germanium-Antimony-Telluride Phase Change Materials for Non-Volatile Memories and Optical Modulators. *Appl. Sci.* **2019**, *9*, 530. [CrossRef]
6. Bruns, G.; Merkelbach, P.; Schlockermann, C.; Salinga, M.; Wuttig, M.; Happ, T.D.; Philipp, J.B.; Kund, M. Nanosecond Switching in GeTe Phase Change Memory Cells. *Appl. Phys. Lett.* **2009**, *95*, 043108. [CrossRef]
7. Loke, D.; Lee, T.H.; Wang, W.J.; Shi, L.P.; Zhao, R.; Yeo, Y.C.; Chong, T.C.; Elliott, S.R. Breaking the Speed Limits of Phase-Change Memory. *Science* **2012**, *336*, 1566–1569. [CrossRef]
8. Ielmini, D.; Wong, H.-S.P. In-Memory Computing with Resistive Switching Devices. *Nat. Electron.* **2018**, *1*, 333–343. [CrossRef]
9. Islam, R.; Li, H.; Chen, P.-Y.; Wan, W.; Chen, H.-Y.; Gao, B.; Wu, H.; Yu, S.; Saraswat, K.; Philip Wong, H.-S. Device and Materials Requirements for Neuromorphic Computing. *J. Phys. Appl. Phys.* **2019**, *52*, 113001. [CrossRef]
10. Spiga, S.; Sebastian, A.; Querlioz, D.; Rajendran, B. Memristive Devices for Brain-Inspired Computing, 1st ed.; Woodhead Publishing Series in Electronic and Optical Materials Series. Available online: https://www.elsevier.com/books/memristive-devices-for-brain-inspired-computing/spiga/978-0-08-102782-0 (accessed on 24 May 2021).
11. Redaelli, A. Self-Consistent Numerical Model. In *Phase Change Memory*; Redaelli, A., Ed.; Springer International Publishing: Cham, Switzerland, 2018; pp. 65–88. [CrossRef]
12. Zhou, X.; Xia, M.; Rao, F.; Wu, L.; Li, X.; Song, Z.; Feng, S.; Sun, H. Understanding Phase-Change Behaviors of Carbon-Doped $Ge_2Sb_2Te_5$ for Phase-Change Memory Application. *ACS Appl. Mater. Interfaces* **2014**, *6*, 14207–14214. [CrossRef]
13. Jeong, T.H.; Seo, H.; Lee, K.L.; Choi, S.M.; Kim, S.J.; Kim, S.Y. Study of Oxygen-Doped GeSbTe Film and Its Effect as an Interface Layer on the Recording Properties in the Blue Wavelength. *Jpn. J. Appl. Phys.* **2001**, *40*, 1609–1612. [CrossRef]
14. Wang, K.; Wamwangi, D.; Ziegler, S.; Steimer, C.; Wuttig, M. Influence of Bi Doping upon the Phase Change Characteristics of $Ge_2Sb_2Te_5$. *J. Appl. Phys.* **2004**, *96*, 5557–5562. [CrossRef]
15. Choi, K.-J.; Yoon, S.-M.; Lee, N.-Y.; Lee, S.-Y.; Park, Y.-S.; Yu, B.-G.; Ryu, S.-O. The Effect of Antimony-Doping on $Ge_2Sb_2Te_5$, a Phase Change Material. *Thin Solid Films* **2008**, *516*, 8810–8812. [CrossRef]
16. Kim, K.-H.; Chung, J.-G.; Kyoung, Y.K.; Park, J.-C.; Choi, S.-J. Phase-Change Characteristics of Nitrogen-Doped $Ge_2Sb_2Te_5$ Films during Annealing Process. *J. Mater. Sci. Mater. Electron.* **2011**, *22*, 52–55. [CrossRef]
17. Navarro, G.; Sousa, V.; Noe, P.; Castellani, N.; Coue, M.; Kluge, J.; Kiouseloglou, A.; Sabbione, C.; Persico, A.; Roule, A.; et al. N-Doping Impact in Optimized Ge-Rich Materials Based Phase-Change Memory. In Proceedings of the 2016 IEEE 8th International Memory Workshop (IMW), Paris, France, 15–18 May 2016; pp. 1–4. [CrossRef]

18. Luong, M.A.; Wen, D.; Rahier, E.; Ratel Ramond, N.; Pecassou, B.; Le Friec, Y.; Benoit, D.; Claverie, A. Impact of Nitrogen on the Crystallization and Microstructure of Ge-Rich GeSbTe Alloys. *Phys. Status Solidi RRL Rapid Res. Lett.* **2021**, *15*, 2000443. [CrossRef]
19. Luong, M.A.; Agati, M.; Ratel Ramond, N.; Grisolia, J.; Le Friec, Y.; Benoit, D.; Claverie, A. On Some Unique Specificities of Ge-Rich GeSbTe Phase-Change Material Alloys for Nonvolatile Embedded-Memory Applications. *Phys. Status Solidi Rapid Res. Lett.* **2021**, *15*, 2000471. [CrossRef]
20. Shelby, R.M.; Raoux, S. Crystallization Dynamics of Nitrogen-Doped $Ge_2Sb_2Te_5$. *J. Appl. Phys.* **2009**, *105*, 104902. [CrossRef]
21. Yu, X.; Zhao, Y.; Li, C.; Hu, C.; Ma, L.; Fan, S.; Zhao, Y.; Min, N.; Tao, S.; Wang, Y. Improved Multi-Level Data Storage Properties of Germanium-Antimony-Tellurium Films by Nitrogen Doping. *Scr. Mater.* **2017**, *141*, 120–124. [CrossRef]
22. Kim, Y.; Baeck, J.H.; Cho, M.-H.; Jeong, E.J.; Ko, D.-H. Effects of N_2^+ Ion Implantation on Phase Transition in $Ge_2Sb_2Te_5$ Films. *J. Appl. Phys.* **2006**, *100*, 083502. [CrossRef]
23. Privitera, S.; Rimini, E.; Zonca, R. Amorphous-to-Crystal Transition of Nitrogen- and Oxygen-Doped $Ge_2Sb_2Te_5$ Films Studied by In Situ Resistance Measurements. *Appl. Phys. Lett.* **2004**, *85*, 3044–3046. [CrossRef]
24. Lai, Y.; Qiao, B.; Feng, J.; Ling, Y.; Lai, L.; Lin, Y.; Tang, T.; Cai, B.; Chen, B. Nitrogen-Doped $Ge_2Sb_2Te_5$ Films for Nonvolatile Memory. *J. Electron. Mater.* **2005**, *34*, 176–181. [CrossRef]
25. Wimmer, M.; Kaes, M.; Dellen, C.; Salinga, M. Role of Activation Energy in Resistance Drift of Amorphous Phase Change Materials. *Front. Phys.* **2014**, *2*. [CrossRef]
26. Noé, P.; Sabbione, C.; Castellani, N.; Veux, G.; Navarro, G.; Sousa, V.; Hippert, F.; d'Acapito, F. Structural Change with the Resistance Drift Phenomenon in Amorphous GeTe Phase Change Materials' Thin Films. *J. Phys. Appl. Phys.* **2016**, *49*, 035305. [CrossRef]
27. Park, I.-M.; Cho, J.-Y.; Yang, T.-Y.; Park, E.S.; Joo, Y.-C. Thermomechanical Analysis on the Phase Stability of Nitrogen-Doped Amorphous $Ge_2Sb_2Te_5$ Films. *Jpn. J. Appl. Phys.* **2011**, *50*, 061201. [CrossRef]
28. Jeong, T.H.; Kim, M.R.; Seo, H.; Park, J.W.; Yeon, C. Crystal Structure and Microstructure of Nitrogen-Doped $Ge_2Sb_2Te_5$ Thin Film. *Jpn. J. Appl. Phys.* **2000**, *39*, 2775–2779. [CrossRef]
29. Kooi, B.J.; Groot, W.M.G.; De Hosson, J.T.M. In Situ Transmission Electron Microscopy Study of the Crystallization of $Ge_2Sb_2Te_5$. *J. Appl. Phys.* **2004**, *95*, 924–932. [CrossRef]
30. Liu, B.; Song, Z.; Zhang, T.; Xia, J.; Feng, S.; Chen, B. Effect of N-Implantation on the Structural and Electrical Characteristics of $Ge_2Sb_2Te_5$ Phase Change Film. *Thin Solid Films* **2005**, *478*, 49–55. [CrossRef]
31. Kim, K.-H.; Park, J.-C.; Lee, J.-H.; Chung, J.-G.; Heo, S.; Choi, S.-J. Nitrogen-Doping Effect on $Ge_2Sb_2Te_5$ Chalcogenide Alloy Films during Annealing. *Jpn. J. Appl. Phys.* **2010**, *49*, 101201. [CrossRef]
32. Kim, Y.; Jeong, K.; Cho, M.-H.; Hwang, U.; Jeong, H.S.; Kim, K. Changes in the Electronic Structures and Optical Band Gap of Ge2Sb2Te5 and N-Doped $Ge_2Sb_2Te_5$ during Phase Transition. *Appl. Phys. Lett.* **2007**, *90*, 171920. [CrossRef]
33. Caravati, S.; Colleoni, D.; Mazzarello, R.; Kühne, T.D.; Krack, M.; Bernasconi, M.; Parrinello, M. First-Principles Study of Nitrogen Doping in Cubic and Amorphous $Ge_2Sb_2Te_5$. *J. Phys. Condens. Matter* **2011**, *23*, 265801. [CrossRef]
34. Kojima, R.; Okabayashi, S.; Kashihara, T.; Horai, K.; Matsunaga, T.; Ohno, E.; Yamada, N.; Ohta, T. Nitrogen Doping Effect on Phase Change Optical Disks. *Jpn. J. Appl. Phys.* **1998**, *37*, 2098–2103. [CrossRef]
35. Wang, W.; Loke, D.; Shi, L.; Zhao, R.; Yang, H.; Law, L.-T.; Ng, L.-T.; Lim, K.-G.; Yeo, Y.-C.; Chong, T.-C.; et al. Enabling Universal Memory by Overcoming the Contradictory Speed and Stability Nature of Phase-Change Materials. *Sci. Rep.* **2012**, *2*, 360. [CrossRef]
36. Bert, N.A.; Chaldyshev, V.V.; Cherkashin, N.A.; Nevedomskiy, V.N.; Preobrazhenskii, V.V.; Putyato, M.A.; Semyagin, B.R.; Ushanov, V.I.; Yagovkina, M.A. Metallic AsSb Nanoinclusions Strongly Enriched by Sb in AlGaAsSb Metamaterial. *J. Appl. Phys.* **2019**, *125*, 145106. [CrossRef]
37. 80pm Resolution Reached by the New CEMES-Hitachi I²TEM Microscope. Actualités Next Toulouse, Nano, Mesures Extrêmes, Théorie. Laboratoires Recherche Physique Chimie. Available online: http://www.next-toulouse.eu/news/80pm-resolution-reached-by-the-new-cemes-hitachi-i2tem-microscope-en (accessed on 24 May 2021).
38. Agati, M.; Gay, C.; Benoit, D.; Claverie, A. Effects of Surface Oxidation on the Crystallization Characteristics of Ge-Rich Ge-Sb-Te Alloys Thin Films. *Appl. Surf. Sci.* **2020**, *518*, 146227. [CrossRef]
39. James Ziegler—SRIM & TRIM. Available online: http://www.srim.org/ (accessed on 24 May 2021).
40. Fantini, A.; Sousa, V.; Perniola, L.; Gourvest, E.; Bastien, J.; Maitrejean, S.; Braga, S.; Pashkov, N.; Bastard, A.; Hyot, B.; et al. N-Doped GeTe as Performance Booster for Embedded Phase-Change Memories. In Proceedings of the 2010 International Electron Devices Meeting, San Francisco, CA, USA, 6–10 December 2010; pp. 29.1.1–29.1.4. [CrossRef]
41. Burr, G.W.; Breitwisch, M.J.; Franceschini, M.; Garetto, D.; Gopalakrishnan, K.; Jackson, B.; Kurdi, B.; Lam, C.; Lastras, L.A.; Padilla, A.; et al. Phase Change Memory Technology. *J. Vac. Sci. Technol. B Nanotechnol. Microelectron. Mater. Process. Meas. Phenom.* **2010**, *28*, 223–262. [CrossRef]
42. Kalb, J.A. Crystallization Kinetics. In *Phase Change Materials*; Raoux, S., Wuttig, M., Eds.; Springer: Boston, MA, USA, 2009. [CrossRef]
43. Redaelli, A. An Introduction on Phase-Change Memories. In *Phase Change Memory*; Redaelli, A., Ed.; Springer International Publishing: Cham, Switzerland, 2018; pp. 1–10. [CrossRef]
44. Lee, T.H.; Loke, D.; Elliott, S.R. Microscopic Mechanism of Doping-Induced Kinetically Constrained Crystallization in Phase-Change Materials. *Adv. Mater.* **2015**, *27*, 5477–5483. [CrossRef]

45. Song, S.A.; Zhang, W.; Sik Jeong, H.; Kim, J.-G.; Kim, Y.-J. In Situ Dynamic HR-TEM and EELS Study on Phase Transitions of Ge$_2$Sb$_2$Te$_5$ Chalcogenides. *Ultramicroscopy* **2008**, *108*, 1408–1419. [CrossRef]
46. Hÿtch, M.J. Analysis of Variations in Structure from High Resolution Electron Microscope Images by Combining Real Space and Fourier Space Information. *Microsc. Microanal. Microstruct.* **1997**, *8*, 41–57. [CrossRef]
47. Bierwolf, R.; Hohenstein, M.; Phillipp, F.; Brandt, O.; Crook, G.E.; Ploog, K. Direct Measurement of Local Lattice Distortions in Strained Layer Structures by HREM. *Ultramicroscopy* **1993**, *49*, 273–285. [CrossRef]
48. Usuda, K.; Numata, T.; Irisawa, T.; Hirashita, N.; Takagi, S. Strain Characterization in SOI and Strained-Si on SGOI MOSFET Channel Using Nano-Beam Electron Diffraction (NBD). *Mater. Sci. Eng. B* **2005**, *124–125*, 143–147. [CrossRef]
49. Hÿtch, M.; Houdellier, F.; Hüe, F.; Snoeck, E. Nanoscale Holographic Interferometry for Strain Measurements in Electronic Devices. *Nature* **2008**, *453*, 1086–1089. [CrossRef]
50. Louiset, A.; Schamm-Chardon, S.; Kononchuk, O.; Cherkashin, N. Reconstruction of Depth Resolved Strain Tensor in Off-Axis Single Crystals: Application to H$^+$ Ions Implanted LiTaO$_3$. *Appl. Phys. Lett.* **2021**, *118*, 082903. [CrossRef]
51. Kim, K.; Park, J.-C.; Chung, J.-G.; Song, S.A.; Jung, M.-C.; Lee, Y.M.; Shin, H.-J.; Kuh, B.; Ha, Y.; Noh, J.-S. Observation of Molecular Nitrogen in N-doped Ge$_2$Sb$_2$Te$_5$. *Appl. Phys. Lett.* **2006**, *89*, 243520. [CrossRef]

Article

Tailoring the Structural and Optical Properties of Germanium Telluride Phase-Change Materials by Indium Incorporation

Xudong Wang [1], Xueyang Shen [1], Suyang Sun [1] and Wei Zhang [1,2,*]

[1] Center for Alloy Innovation and Design (CAID), State Key Laboratory for Mechanical Behavior of Materials, Xi'an Jiaotong University, Xi'an 710049, China; xudong.wang@stu.xjtu.edu.cn (X.W.); v32267209@stu.xjtu.edu.cn (X.S.); sy.sun@stu.xjtu.edu.cn (S.S.)
[2] Pazhou Lab, Pengcheng National Laboratory in Guangzhou, Guangzhou 510320, China
* Correspondence: wzhang0@mail.xjtu.edu.cn

Abstract: Chalcogenide phase-change materials (PCMs) based random access memory (PCRAM) enter the global memory market as storage-class memory (SCM), holding great promise for future neuro-inspired computing and non-volatile photonic applications. The thermal stability of the amorphous phase of PCMs is a demanding property requiring further improvement. In this work, we focus on indium, an alloying ingredient extensively exploited in PCMs. Starting from the prototype GeTe alloy, we incorporated indium to form three typical compositions along the InTe-GeTe tie line: $InGe_3Te_4$, $InGeTe_2$ and In_3GeTe_4. The evolution of structural details, and the optical properties of the three In-Ge-Te alloys in amorphous and crystalline form, was thoroughly analyzed via ab initio calculations. This study proposes a chemical composition possessing both improved thermal stability and sizable optical contrast for PCM-based non-volatile photonic applications.

Keywords: phase change materials; amorphous phase; germanium telluride; indium alloying; optical contrast

1. Introduction

Non-volatile memory (NVM) is a rising technology that allows for high-density data storage and fast data processing [1–6]. Phase-change materials (PCMs)-based random access memory (PCRAM) is a leading NVM candidate with successful stand-alone memory products such as Intel Optane. By improving its thermal stability, PCRAM is also a promising candidate for embedded memory [7,8]. As announced by STMicroelectronics, PCRAM will be used as embedded memory, replacing Flash memory, for their future microcontroller units (MCU) for the automotive industry [7]. Moreover, PCRAM is also being exploited for more advanced applications, including neuro-inspired computing [9–15], stochasticity-based computing [16,17], flexible electronics [18], optical displays [19–21], all-optical computers [22–26], low-loss optical modulators [27], metasurfaces [28–32] and others [33,34].

PCMs can be switched rapidly and reversibly between their amorphous and crystalline phases via Joule heating induced by electrical or optical pulses [1,35]. The notable contrast in either electrical resistivity or optical reflectivity between each phase is utilized to encode digital information [1]. Several demanding requirements, such as high programming speed, good thermal stability, low power consumption, stable property contrast window and long cycling endurance, have to be well satisfied for high-performance PCRAM. Germanium chalcogenides, in particular, GeTe and GeTe-Sb_2Te_3 pseudo-binary compounds (GST), especially $Ge_2Sb_2Te_5$ [36], are one of the most successful material families that could meet these challenging requirements simultaneously. Doping and alloying are frequently used to tailor the material properties for faster speed and/or better retention temperature, targeting different application scenarios [37–44].

For decades, indium has been an important alloying element used in rewritable optical data storage products [45]. The flagship PCM is AgInSbTe [45–48]. Recently, indium-alloyed GeTe [49–52] and GST [53,54] were reported, and their enhanced amorphous stability makes them suitable candidates for high-temperature PCRAM applications. In addition, indium forms a unique PCM In_3SbTe_2 [55–59] that exhibits metallic behavior in its crystalline phase, but semiconducting behaviors in its amorphous phase, in contrast to conventional PCMs, which remain semiconducting during memory programming. It has been suggested that even InTe could also be a potential PCM for non-volatile electronics [53,60]. In this work, we focus on the InTe-GeTe (IGT) tie line, in particular, the three stoichiometric compositions, namely $InGe_3Te_4$, $InGeTe_2$ and In_3GeTe_4. By performing thorough ab initio calculations and chemical bonding analyses, we elucidate the role of indium in altering the structural and optical properties of GeTe.

2. Computational Details

We performed ab initio molecular dynamics (AIMD) simulations based on density functional theory (DFT) to generate melt-quenched amorphous structures [61]. The second-generation Car–Parrinello method [62] as implemented in CP2K package [63] was employed along with Perdew–Burke–Ernzerhof (PBE) functional [64] and the Goedecker pseudopotentials [65]. The canonical NVT ensemble was used and the time step was set at 2 fs. Vienna Ab-initio Simulation Package (VASP) [66] was employed to relax the amorphous structures and crystalline counterparts, prior to the calculations of electronic structure and optical response. For VASP calculations, we applied the PBE functional and projector augmented-wave (PAW) pseudopotentials [67]. The energy cutoff for plane waves was set at 500 eV. Chemical bonding analyses were conducted with the LOBSTER code [68–70]. Crystal orbital Hamilton populations (COHP) were applied to separate the covalent interactions into bonding (positive −COHP) and antibonding (negative −COHP) contributions. Bader charges were calculated to evaluate the atomic charge transfer in the structures [71]. Frequency-dependent dielectric matrix was calculated within the independent-particle approximation without considering local field effects and many body effects, which proved to be adequate to account for the optical contrast between crystalline and amorphous PCMs [72–74]. The absorption $\alpha(\omega)$ and reflectivity $R(\omega)$ can be calculated from the dielectric functions [75]:

$$\alpha(\omega) = \frac{\sqrt{2}\omega}{c} \left(\sqrt{\varepsilon_1^2 + \varepsilon_2^2} - \varepsilon_1 \right)^{\frac{1}{2}} \quad (1)$$

$$R(\omega) = \frac{(n-1)^2 + k^2}{(n+1)^2 + k^2} \quad (2)$$

where ε_1 and ε_2 are the real and imaginary parts of the dielectric function. n and k are the refractive index and extinction coefficient, which can be calculated from the dielectric functions:

$$n(\omega) = \left(\frac{\sqrt{\varepsilon_1^2 + \varepsilon_2^2} + \varepsilon_1}{2} \right)^{\frac{1}{2}} \quad (3)$$

$$k(\omega) = \left(\frac{\sqrt{\varepsilon_1^2 + \varepsilon_2^2} - \varepsilon_1}{2} \right)^{\frac{1}{2}} \quad (4)$$

Van der Waals correction based on the Grimme's D3 method was considered in all AIMD and DFT calculations [76,77]. All the electronic structures, chemical bonding and optical properties were calculated using relaxed structures at zero K with VASP. For standard calculations, only gamma point was used to sample the Brillouin zone of the supercell models, while a 3 × 3 × 3 k-point mesh was used to converge the optical response calculations. For statistics, we built three crystalline and three amorphous models for each composition.

3. Results and Discussion

As reported in Ref. [78], a single-phase rock-salt structure was obtained over a wide compositional range of 8–75 mole% InTe in IGT at ambient conditions, in which Te atoms occupied one sublattice, while Ge and In atoms shared the other one. The three IGT compositions considered in this work, namely $InGe_3Te_4$, $InGeTe_2$ and In_3GeTe_4, fall in this compositional range, and were expected to take the rock-salt structure. To account for the compositional disorder of Ge and In atoms on the cation-like sublattice, we built $3 \times 3 \times 3$ supercells (216 atoms in total) and distributed Ge and In atoms using a quasi-random number generator. Three independent models were considered for each composition. Each supercell model was fully relaxed with respect to both atomic coordinates and cell volume by DFT calculations. The relaxed cell-edge of crystalline (c-) $InGe_3Te_4$, $InGeTe_2$ and In_3GeTe_4 is 18.21, 18.28 and 18.55 Å, respectively. The corresponding unit cell lattice parameters, 6.07, 6.09 and 6.18 Å, are in good agreement with experimental values (5.97, 6.00 and 6.06 Å) [78]. The relaxed structure of c-$InGeTe_2$ is shown in Figure 1a and the structures of the other two compositions are shown in Figure S1.

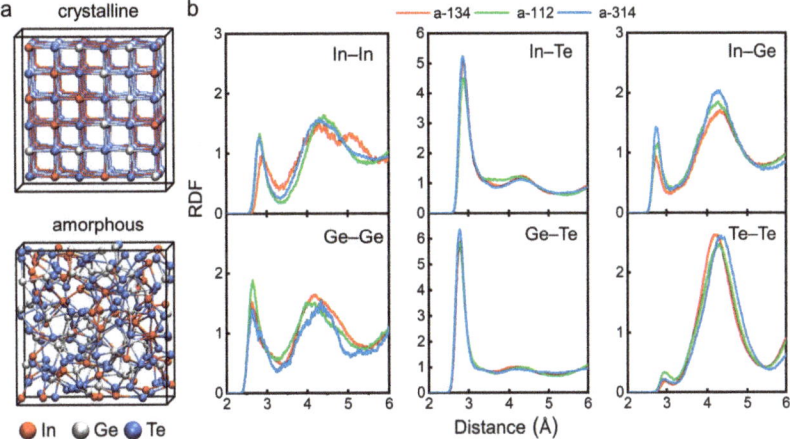

Figure 1. (a) Atomic structures for crystalline and amorphous $InGeTe_2$. Red, silver and blue spheres represent In, Ge and Te atoms, respectively. (b) Partial radial distribution functions (RDFs) for each atomic pair in the three amorphous compounds. The "a-134", "a-112" and "a-314" represent amorphous $InGe_3Te_4$, $InGeTe_2$ and In_3GeTe_4, respectively.

The relaxed crystalline supercells were then used to generate amorphous (a-) models following a melt–quench protocol [61]. The supercell models were quickly heated to a very high temperature to remove the crystalline order. After randomization at 3000 K for 15 ps, the models were quenched down to and equilibrated at 1200 K, above the melting point of IGT alloys (∼550–750 °C) [49] for 30 ps. Amorphous models were then generated by quenching the liquids down to 300 K with a cooling rate of 12.5 K/ps. During this quenching process, we stopped the simulation after every 100 K, and the simulation box size was increased to reduce the internal stress. Within each temperature window, one NVT calculation was performed using a fixed box size. The model was equilibrated at 300 K for 30 ps. This density value was then used to generate two additional melt-quenched amorphous models for each composition. All three amorphous models showed consistently low pressure values below 3 kbar. The obtained cell edges of amorphous $InGe_3Te_4$, $InGeTe_2$ and In_3GeTe_4 are 18.90, 19.08 and 19.32 Å, respectively. This increase in the cell edge of the amorphous phase is consistent with the trend observed in their crystalline counterparts.

Further optimization of the internal stress or the use of the NPT ensemble for melt-quench simulations could potentially lead to some numerical differences in the mass density, but is not expected to alter the amorphous structures much.

The amorphous structure of InGeTe$_2$ is shown in Figure 1a and the snapshots of the other two compositions are in Figure S1. The partial radial distribution functions (RDFs) of each atomic pair in the three amorphous compounds are shown in Figure 1b. The peak positions of the heteropolar bonds In–Te (2.87 Å) and Ge–Te (2.78 Å) are not varied with composition, whereas the homopolar or "wrong" bonds show small shifts. As developed in our previous work [79], the "bond-weighted distribution function (BWDF)" provides direct information on the length of chemical bonds in amorphous IGT alloys (Figure S2). Despite the change in chemical composition, the bond length shows very similar values in amorphous IGT alloys, i.e., Ge–Te 3.20 Å, In–Te 3.40 Å, Ge–Ge 3.20 Å, In–In 3.25 Å and Ge–In 3.40 Å. In all three amorphous IGT alloys, Te–Te shows mostly antibonding interactions. These bond length values are used as cutoffs for the interatomic distance for the following structural analysis.

The angle distribution function (ADF) of the three amorphous structures (Figure 2a) shows that In and Ge atoms mainly form local motifs with the central bond angles ranging from 90° to 109.5°, which corresponds to the bond angles in octahedral and tetrahedral motifs, respectively. As the concentration of indium increases, the ADF peak for indium atoms clearly shifts toward 109.5°, implying an increase in indium-centered tetrahedral motifs. We used bond order parameter q [80] to quantify the fraction of tetrahedral motifs in amorphous IGT alloys. Such parameters are frequently used for the structural analysis of amorphous PCMs [81–83]. As shown in Figure 2b, as indium concentration increases, the total fraction of In- and Ge-centered tetrahedral motifs (short as tetra-In and tetra-Ge) increases from 31.6% (a-InGe$_3$Te$_4$), 33.5% (a-InGeTe$_2$) to 42.4% (a-In$_3$GeTe$_4$) in relation to the total number of In and Ge atoms. Specifically, the fraction of tetra-In increases from 10.1%, 18.6% to 34.6%, while the fraction of tetra-Ge decreases from 21.5%, 14.9% to 7.8%. The fraction of tetrahedrons for a-InGeTe$_2$ is smaller than that reported in previous work (38 % in total, with 31.5% for tetra-In and 6.5% for tetra-Ge) [50], due to the different choice of cutoff values for the interatomic distance and the deviation in calculated density values (vdW interactions were included in the current work). The ratios of tetrahedral motifs in amorphous IGT alloys are all higher than that of their parent phase—GeTe, where 25–30% tetrahedral atoms are typically found in the rapidly quenched amorphous phase [84–86].

In addition to the increase in the total number of tetrahedral motifs from a-InGe$_3$Te$_4$ to a-In$_3$GeTe$_4$, the local bonding configuration also shows a major difference. Despite the change in chemical compositions, nearly all the tetra-Ge atoms are bonded with at least one Ge or In atom, while the majority of tetra-In atoms are heteropolar-bonded in the three amorphous IGT alloys. To quantify the role of "wrong" bonds, we carried out a projected COHP (pCOHP) analysis. As shown in Figures 3 and S3, the pCOHP of heteropolar-bonded tetra-Ge atoms demonstrates sizable antibonding interactions right below Fermi energy (E_F), and the presence of wrong bonds (including Ge–Ge and Ge–In) largely reduces such antibonding contributions, stabilizing the tetrahedral motifs locally. In contrast, the pCOHP of tetra-In atoms with and without wrong bonds mostly shows bonding interactions below E_F. These results are consistent with the bonding configuration in the two parent phases, a-GeTe [79] and a-InTe [83], though In–Ge bonds are present in all three amorphous IGT alloys. Since indium atoms do not require homopolar bonds to stabilize tetrahedral motifs, the ratio of tetrahedral units is increased with the indium concentration. As compared to their crystalline counterparts, where all In and Ge atoms are octahedrally bonded, the enlarged structural deviation will enhance the thermal stability of amorphous IGT alloys. This observation is consistent with experimental findings, as the crystallization temperature T_x of doped InGeTe$_2$ [49] and undoped InTe thin films [60] is increased to ∼276 and ∼300 °C, respectively, as compared to that of GeTe T_x ∼190 °C [87].

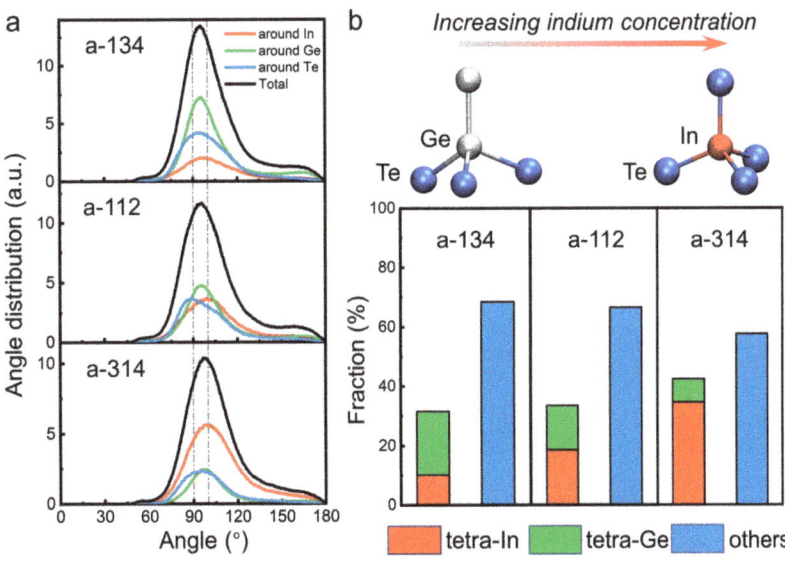

Figure 2. (**a**) Angle distribution functions (ADFs) for the amorphous structures of the three In-Ge-Te compositions. The central bond angles for perfect tetrahedron (109.5°) and octahedron (90°) are marked as dashed lines in ADF plots. (**b**) Ratios of tetrahedral motifs in the three amorphous structures. Typical Ge- and In-centered tetrahedral motifs (short as tetra-In and tetra-Ge) in the amorphous structures are shown on the top panel.

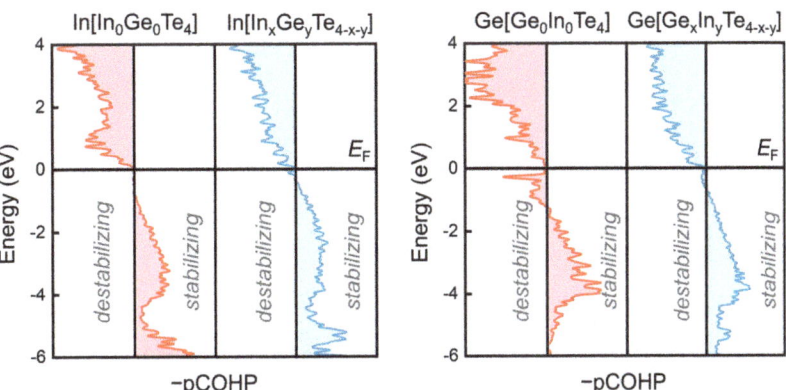

Figure 3. Projected COHP (pCOHP) for tetra-In and tetra-Ge motifs in a-InGeTe$_2$. Tetrahedral motifs are classified as the ones with only heteropolar bonds, denoted as In[In$_0$Ge$_0$Te$_4$] and Ge[Ge$_0$In$_0$Te$_4$], and the others with at least one wrong bond indicated as In[In$_x$Ge$_y$Te$_{4-x-y}$] and Ge[Ge$_x$In$_y$Te$_{4-x-y}$] (x or $y \geq 1$, $x + y \leq 4$).

The calculated density of states (DOS) of the three IGT alloys in both crystalline and amorphous forms are shown in Figure 4a,b. Regarding the crystalline models, the overall DOS profiles are quite similar, and all three alloys exhibit metallic features. By contrast, all three amorphous models are narrow-gap semiconductors. Statistical sampling yields consistent results (Figure S4). The large difference in DOS between the crystalline and amorphous IGT results in a wide resistance contrast window for PCRAM applications [49]. The Bader charge analysis (Figure 4c) details larger net charges for In atoms than for Ge atoms due to the difference in electronegativity. The bimodal feature of the charges of indium atoms is consistent with previous work [50], stemming from different local environments of indium atoms. The enlarged charge transfer in amorphous structures increases the probability of long-distance electromigration under the transient electrical field induced by programming pulses [88], which is detrimental to the cycling endurance of devices [89,90]. For RESET operations, the higher the melting temperature T_m, the greater the power consumption. The melting temperature for IGT alloys has a "W" shape profile, according to the InTe-GeTe phase diagram, which shows that $InGeTe_2$ has a higher melting temperature T_m (740 °C) than GeTe and InTe whose T_m are 715 and 688 °C, respectively [49]. Interestingly, the two other compositions, $InGe_3Te_4$ and In_3GeTe_4, demonstrated reduced T_m (645 and 565 °C). Taking into account all these factors for practical applications, we would suggest keeping the IGT composition within the range of $InGeTe_2$ to In_3GeTe_4 for balanced device performance.

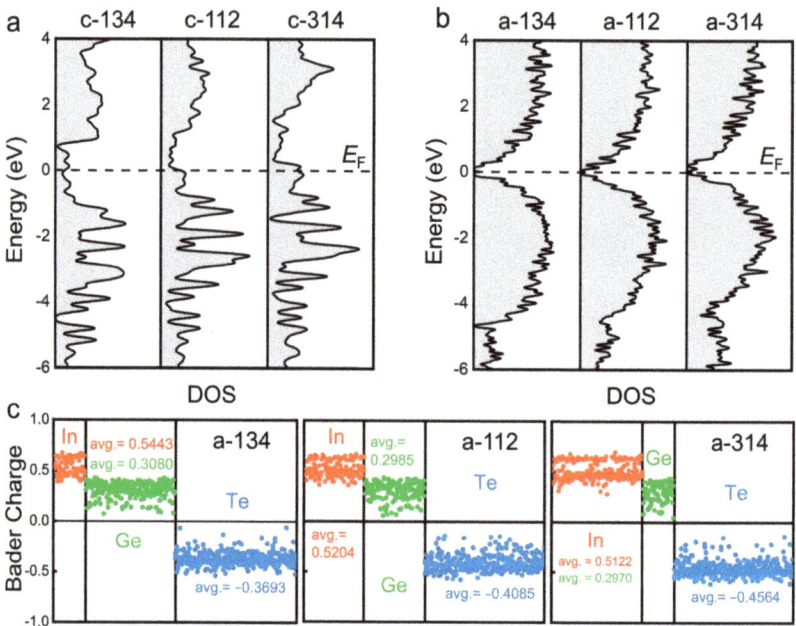

Figure 4. Density of states (DOS) for the (**a**) crystalline and (**b**) amorphous structures of the three In-Ge-Te compositions. The "c-134", "c-112" and "c-314" represent crystalline $InGe_3Te_4$, $InGeTe_2$ and In_3GeTe_4, respectively. (**c**) Bader charges (in electrons per atom) of all atoms in the three amorphous structures.

The enhanced amorphous stability is also useful for non-volatile photonic applications [91], yet the incorporation of indium makes IGT alloys metallic, which could affect the optical contrast between the amorphous and crystalline phase. The significant contrast of ~30% in the optical reflectivity of PCMs stems from a fundamental change in bonding nature from covalent to metavalent bonding (MVB) upon crystallization [92–99]. However, in comparison with GeTe and GST, which have three p electrons per site (a key feature of MVB), InTe has a deficient number of p electrons, turning the rock-salt phase from semiconducting to metallic. As a result, MVB in IGT alloys is expected to be weakened.

For verification, we carried out optical response calculations using the relaxed crystalline and amorphous IGT structures. We focused on the spectrum range from 400 to 1600 nm, covering both the visible light region (~400 nm to 800 nm) for optical displays [19–21] and the telecom wavelength bands (~1500 to 1600 nm) for silicon-waveguide-integrated photonic applications [23–26]. As shown in Figure 5a, the optical absorption and reflectivity profiles vary slightly with the chemical compositions in the amorphous phase, while strong changes are found in the crystalline phase. For InGe$_3$Te$_4$ and InGeTe$_2$, sizable contrast in reflectivity between the crystalline and amorphous phases is observed over the whole spectrum, with an average value ~20%. However, for In$_3$GeTe$_4$, the contrast in reflectivity nearly vanishes at around ~900 nm, and it becomes greater below 700 nm, or above 1000 nm.

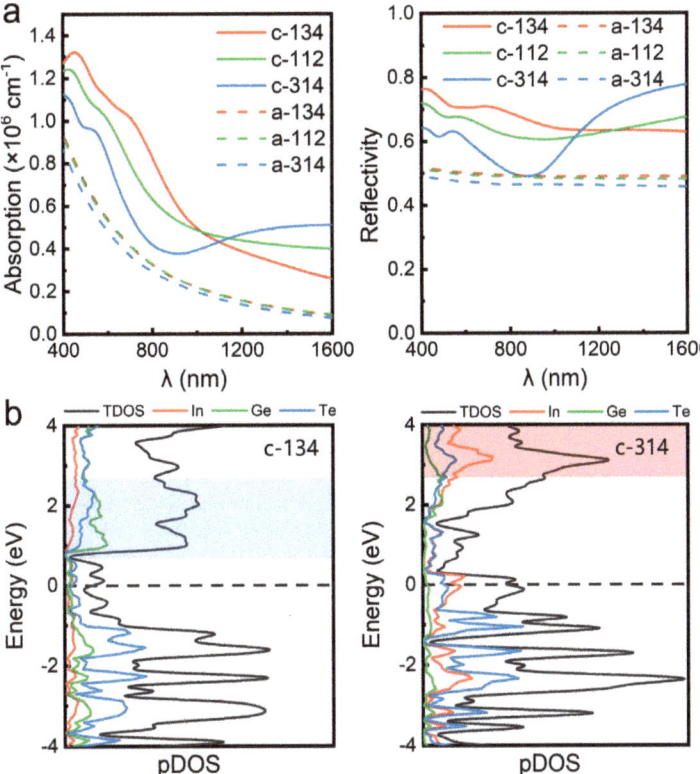

Figure 5. (**a**) Optical absorption and reflectivity of crystalline and amorphous phase of the three In-Ge-Te compositions. (**b**) Projected DOS (pDOS) of c-134 and c-314. Blue and red shaded areas highlight the peak regions of DOS above E_F in c-134 and c-314, respectively.

The three IGT crystals have a crossover at ~1100 nm. For long wavelength or small photon energy (below ~1 eV) regions, the optical excitation is mainly determined by states near E_F. With increased DOS near E_F by heavier indium alloying (Figure 4a), the absorption in the long wavelength region is enhanced. However, for the short wavelength or large photon energy (above ~2 eV) region, the electronic states of a wider energy range would participate in the optical excitation. According to the projected DOS (Figure 5b), the valence states below E_F are mainly contributed by Te atoms for all three crystals. However, indium alloying shifts the DOS peak in the conduction band to a higher energy range. Since fewer excited states of low energy could contribute to short wavelength excitation, c-In$_3$GeTe$_4$ shows the smallest absorption and reflectivity in the short wavelength region (Figure 5a).

We note that our optical calculations were performed using DFT-PBE functional with an independent-particle approximation, excluding local field effects and many body effects. Therefore, the absolute values of optical profiles could vary if more advanced methods are employed. Nevertheless, the observation of weakened optical contrast due to heavier indium alloying should remain valid. Taking into account the enhanced crystallization temperature and the reduced melting temperature, we predict an optimal IGT composition within the ranges of $InGeTe_2$ and In_3GeTe_4 for high-performance non-volatile photonics. To the best of our knowledge, thorough optical measurements of IGT alloys are still lacking. Therefore, we anticipate future experiments exploring the suitability of IGT alloys for optical and photonic PCM applications.

4. Conclusions

In summary, we have carried out systematic ab initio calculations for three typical compositions of indium incorporated GeTe compounds, $InGe_3Te_4$, $InGeTe_2$ and In_3GeTe_4, to elucidate the evolution of structural and optical properties along the InTe-GeTe tie line. Upon indium alloying, the crystalline phase of all the three alloys turns metallic, while their amorphous counterparts all show semiconducting features with narrow band gaps. This stark contrast in the electronic structure guarantees a large resistance window between amorphous and crystalline In-Ge-Te alloys for electrical PCRAM. Yet, too much indium should be avoided, because the stronger charge transfer could be harmful to cycling endurance due to electromigration. Regarding optical properties, both $InGe_3Te_4$ and $InGeTe_2$ show sizable optical contrast between the crystalline and amorphous phases in the spectrum range from 400 nm to 1600 nm, covering both visible-light and telecom bands. Meanwhile, In_3GeTe_4 shows a less robust contrast window, due to weakened MVB. Moreover, the increased indium concentration enlarges the ratio of tetrahedral motifs in the amorphous phase and consequently increases the structural barrier for crystallization. The InTe-GeTe phase diagram establishes that the melting temperature reaches minimum around In_3GeTe_4, indicating lowest power consumption for melt–quench amorphization. Taking all these factors into account, we suggest that the optimal chemical composition for In-Ge-Te alloys should be located in the range between $InGeTe_2$ and In_3GeTe_4, which could result in the most balanced device performance for PCM-based non-volatile electronic and photonic applications. Our work should serve as a stimulus for further investigations into indium-incorporated PCMs.

Supplementary Materials: The following are available online at https://www.mdpi.com/article/10.3390/nano11113029/s1. Figure S1: Atomic structures of crystalline and amorphous $InGe_3Te_4$ and In_3GeTe_4 (denoted as c/a-134 and 314); Figure S2: Bond population (left panel) and bond weighted distribution functions BWDFs (right panel) for each interatomic pair in the three amorphous compounds. The crossover values from positive to negative in the BWDFs represent cutoff values for bonding interactions, which are used for structural analysis. Further technical details about BWDF can be found in [79] of the main text; Figure S3: Projected COHP (pCOHP) for tetra-In and tetra-Ge motifs in a-$InGe_3Te_4$ and a-In_3GeTe_4. Tetra-hedral motifs are grouped as the ones with only heteropolar bonds, denoted as $In[In_0Ge_0Te_4]$ and $Ge[Ge_0In_0Te_4]$, and the others with at least one wrong bond indicated as $In[In_xGe_yTe_{4-x-y}]$ and $Ge[Ge_xIn_yTe_{4-x-y}]$ (x or y ≥ 1, x + y ≤ 4). In the a-314 structure, $Ge[Ge_0In_0Te_4]$ motif is absent; Figure S4: Density of states (DOS) for crystalline and amorphous structures of the three IGT compositions. Three models for each composition were built, which show consistent results.

Author Contributions: Investigation, visualization, X.W., X.S. and S.S.; writing—original draft preparation, X.W.; conceptualization, funding acquisition, writing—review and editing, W.Z. All authors have read and agreed to the published version of the manuscript.

Funding: This work was funded by National Natural Science Foundation of China (61774123) and 111 Project 2.0 (BP2018008).

Data Availability Statement: The data presented in this study are available on request from the corresponding author.

Acknowledgments: The authors acknowledge the support by the HPC platform of Xi'an Jiaotong University and the International Joint Laboratory for Micro/Nano Manufacturing and Measurement Technologies of Xi'an Jiaotong University.

Conflicts of Interest: The authors declare no conflict of interest.

References

1. Wuttig, M.; Yamada, N. Phase-change materials for rewriteable data storage. *Nat. Mater.* **2007**, *6*, 824–832. [CrossRef]
2. Wong, H.-S.P.; Raoux, S.; Kim, S.B.; Liang, J.; Reifenberg, J.P.; Rajendran, B.; Asheghi, M.; Goodson, K.E. Phase Change Memory. *Proc. IEEE* **2010**, *98*, 2201. [CrossRef]
3. Raoux, S.; Welnic, W.; Ielmini, D. Phase change materials and their application to nonvolatile memories. *Chem. Rev.* **2010**, *110*, 240–267. [CrossRef] [PubMed]
4. Zhang, W.; Ma, E. Unveiling the structural origin to control resistance drift in phase-change memory materials. *Mater. Today* **2020**, *41*, 156–176. [CrossRef]
5. Zhang, W.; Wuttig, M. Phase Change Materials and Superlattices for Non-Volatile Memories. *Phys. Status Solidi RRL* **2019**, *13*, 1900130. [CrossRef]
6. Li, X.-B.; Chen, N.-K.; Wang, X.-P.; Sun, H.-B. Phase-Change Superlattice Materials toward Low Power Consumption and High Density Data Storage: Microscopic Picture, Working Principles, and Optimization. *Adv. Funct. Mater.* **2018**, *28*, 1803380. [CrossRef]
7. Cappelletti, P.; Annunziata, R.; Arnaud, F.; Disegni, F.; Maurelli, A.; Zuliani, P. Phase change memory for automotive grade embedded NVM applications. *J. Phys. D Appl. Phys.* **2020**, *53*, 193002. [CrossRef]
8. Li, X.; Chen, H.; Xie, C.; Cai, D.; Song, S.; Chen, Y.; Lei, Y.; Zhu, M.; Song, Z. Enhancing the Performance of Phase Change Memory for Embedded Applications. *Phys. Status Solidi RRL* **2019**, *13*, 1800558. [CrossRef]
9. Sebastian, A.; Le Gallo, M.; Burr, G.W.; Kim, S.; BrightSky, M.; Eleftheriou, E. Tutorial: Brain-inspired computing using phase-change memory devices. *J. Appl. Phys.* **2018**, *124*, 111101. [CrossRef]
10. Zhang, W.; Mazzarello, R.; Wuttig, M.; Ma, E. Designing crystallization in phase-change materials for universal memory and neuro-inspired computing. *Nat. Rev. Mater.* **2019**, *4*, 150–168. [CrossRef]
11. Ambrogio, S.; Narayanan, P.; Tsai, H.; Shelby, R.M.; Boybat, I.; Burr, G.W. Equivalent-accuracy accelerated neural network training using analogue memory. *Nature* **2018**, *558*, 60–67. [CrossRef]
12. Sebastian, A.; Le Gallo, M.; Khaddam-Aljameh, R.; Eleftheriou, E. Memory devices and applications for in-memory computing. *Nat. Nanotechnol.* **2020**, *15*, 529–544. [CrossRef]
13. Zhu, J.; Zhang, T.; Yang, Y.; Huang, R. A comprehensive review on emerging artificial neuromorphic devices. *Appl. Phys. Rev.* **2020**, *7*, 011312. [CrossRef]
14. Xu, M.; Mai, X.; Lin, J.; Zhang, W.; Li, Y.; He, Y.; Tong, H.; Hou, X.; Zhou, P.; Miao, X. Recent Advances on Neuromorphic Devices Based on Chalcogenide Phase-Change Materials. *Adv. Funct. Mater.* **2020**, *30*, 2003419. [CrossRef]
15. Ding, K.; Wang, J.; Zhou, Y.; Tian, H.; Lu, L.; Mazzarello, R.; Jia, C.; Zhang, W.; Rao, F.; Ma, E. Phase-change heterostructure enables ultralow noise and drift for memory operation. *Science* **2019**, *366*, 210–215. [CrossRef]
16. Lu, Y.; Li, X.; Yan, L.; Zhang, T.; Yang, Y.; Song, Z.; Huang, R. Accelerated Local Training of CNNs by Optimized Direct Feedback Alignment Based on Stochasticity of 4 Mb C-doped Ge2Sb2Te5 PCM Chip in 40 nm Node. In Proceedings of the 2020 IEEE International Electron Devices Meeting, San Francisco, CA, USA, 12–18 December 2020.
17. Lim, D.H.; Wu, S.; Zhao, R.; Lee, J.H.; Jeong, H.; Shi, L. Spontaneous sparse learning for PCM-based memristor neural networks. *Nat. Commun.* **2021**, *12*, 319. [CrossRef]
18. Khan, A.I.; Daus, A.; Islam, R.; Neilson, K.M.; Lee, H.R.; Wong, H.-S.P.; Pop, E. Ultralow switching current density multilevel phase-change memory on a flexible substrate. *Science* **2021**, *373*, 1243–1247. [CrossRef] [PubMed]
19. Hosseini, P.; Wright, C.D.; Bhaskaran, H. An optoelectronic framework enabled by low-dimensional phase-change films. *Nature* **2014**, *511*, 206–211. [CrossRef]
20. Cheng, Z.; Milne, T.; Salter, P.; Kim, J.S.; Humphrey, S.; Booth, M.; Bhaskaran, H. Antimony thin films demonstrate programmable optical nonlinearity. *Sci. Adv.* **2021**, *7*, eabd7097. [CrossRef] [PubMed]
21. Liu, H.; Dong, W.; Wang, H.; Lu, L.; Ruan, Q.; Tan, Y.S.; Simpson, R.E.; Yang, J.K.W. Rewritable color nanoprints in antimony trisulfide films. *Sci. Adv.* **2020**, *6*, eabb7171. [CrossRef]
22. Ovshinsky, S.R. Optical Cognitive Information Processing—A New Field. *Jpn. J. Appl. Phys.* **2004**, *43*, 4695–4699. [CrossRef]
23. Ríos, C.; Stegmaier, M.; Hosseini, P.; Wang, D.; Scherer, T.; Wright, C.D.; Bhaskaran, H.; Pernice, W.H.P. Integrated all-photonic non-volatile multi-level memory. *Nat. Photon.* **2015**, *9*, 725–732. [CrossRef]
24. Cheng, Z.; Ríos, C.; Youngblood, N.; Wright, C.D.; Pernice, W.H.P.; Bhaskaran, H. Device-Level Photonic Memories and Logic Applications Using Phase-Change Materials. *Adv. Mater.* **2018**, *30*, 1802435. [CrossRef]
25. Feldmann, J.; Youngblood, N.; Karpov, M.; Gehring, H.; Li, X.; Stappers, M.; Le Gallo, M.; Fu, X.; Lukashchuk, A.; Raja, A.S.; et al. Parallel convolutional processing using an integrated photonic tensor core. *Nature* **2021**, *589*, 52–58. [CrossRef] [PubMed]
26. Zhang, H.; Zhou, L.; Lu, L.; Xu, J.; Wang, N.; Hu, H.; Rahman, B.M.A.; Zhou, Z.; Chen, J. Miniature Multilevel Optical Memristive Switch Using Phase Change Material. *ACS Photon.* **2019**, *6*, 2205–2212. [CrossRef]

27. Zhang, Y.; Chou, J.B.; Li, J.; Li, H.; Du, Q.; Yadav, A.; Zhou, S.; Shalaginov, M.Y.; Fang, Z.; Zhong, H.; et al. Broadband transparent optical phase change materials for high-performance nonvolatile photonics. *Nat. Commun.* **2019**, *10*, 4279. [CrossRef]
28. Zhang, Y.; Fowler, C.; Liang, J.; Azhar, B.; Shalaginov, M.Y.; Deckoff-Jones, S.; An, S.; Chou, J.B.; Roberts, C.M.; Liberman, V.; et al. Electrically reconfigurable non-volatile metasurface using low-loss optical phase-change material. *Nat. Nanotechnol.* **2021**, *16*, 661–666. [CrossRef] [PubMed]
29. Dong, W.; Qiu, Y.; Zhou, X.; Banas, A.; Banas, K.; Breese, M.B.H.; Cao, T.; Simpson, R.E. Tunable Mid-Infrared Phase-Change Metasurface. *Adv. Opt. Mater.* **2018**, *6*, 1701346. [CrossRef]
30. Leitis, A.; Heßler, A.; Wahl, S.; Wuttig, M.; Taubner, T.; Tittl, A.; Altug, H. All-Dielectric Programmable Huygens' Metasurfaces. *Adv. Funct. Mater.* **2020**, *30*, 1910259. [CrossRef]
31. Wang, Y.; Landreman, P.; Schoen, D.; Okabe, K.; Marshall, A.; Celano, U.; Wong, H.S.P.; Park, J.; Brongersma, M.L. Electrical tuning of phase-change antennas and metasurfaces. *Nat. Nanotechnol.* **2021**, *16*, 667–672. [CrossRef]
32. Tian, J.; Luo, H.; Yang, Y.; Ding, F.; Qu, Y.; Zhao, D.; Qiu, M.; Bozhevolnyi, S.I. Active control of anapole states by structuring the phase-change alloy Ge2Sb2Te5. *Nat. Commun.* **2019**, *10*, 396. [CrossRef] [PubMed]
33. Gholipour, B. The promise of phase-change materials. *Science* **2019**, *366*, 186–187. [CrossRef] [PubMed]
34. Gerislioglu, B.; Bakan, G.; Ahuja, R.; Adam, J.; Mishra, Y.K.; Ahmadivand, A. The role of Ge2Sb2Te5 in enhancing the performance of functional plasmonic devices. *Mater. Today Phys.* **2020**, *12*, 100178. [CrossRef]
35. Bakan, G.; Gerislioglu, B.; Dirisaglik, F.; Jurado, Z.; Sullivan, L.; Dana, A.; Lam, C.; Gokirmak, A.; Silva, H. Extracting the temperature distribution on a phase-change memory cell during crystallization. *J. Appl. Phys.* **2016**, *120*, 164504. [CrossRef]
36. Yamada, N.; Ohno, E.; Nishiuchi, K.; Akahira, N.; Takao, M. Rapid-phase transitions of GeTe-Sb2Te3 pseudobinary amorphous thin films for an optical disk memory. *J. Appl. Phys.* **1991**, *69*, 2849–2856. [CrossRef]
37. Van Pieterson, L.; Lankhorst, M.H.R.; van Schijndel, M.; Kuiper, A.E.T.; Roosen, J.H.J. Phase-change recording materials with a growth-dominated crystallization mechanism: A materials overview. *J. Appl. Phys.* **2005**, *97*, 083520. [CrossRef]
38. Cheng, H.Y.; Hsu, T.H.; Raoux, S.; Wu, J.Y.; Du, P.Y.; Breitwisch, M.; Zhu, Y.; Lai, E.K.; Joseph, E.; Mittal, S.; et al. A High Performance Phase Change Memory with Fast Switching Speed and High Temperature Retention by Engineering the $Ge_xSb_yTe_z$ Phase Change Material. In Proceedings of the 2011 International Electron Devices Meeting, Washington, DC, USA, 5–7 December 2011.
39. Rao, F.; Ding, K.; Zhou, Y.; Zheng, Y.; Xia, M.; Lv, S.; Song, Z.; Feng, S.; Ronneberger, I.; Mazzarello, R.; et al. Reducing the stochasticity of crystal nucleation to enable subnanosecond memory writing. *Science* **2017**, *358*, 1423–1427. [CrossRef]
40. Liu, B.; Li, K.; Liu, W.; Zhou, J.; Wu, L.; Song, Z.; Elliott, S.R.; Sun, Z. Multi-level phase-change memory with ultralow power consumption and resistance drift. *Sci. Bull.* **2021**, *66*, 2217. [CrossRef]
41. Zewdie, G.M.; Zhou, Y.-X.; Sun, L.; Rao, F.; Deringer, V.L.; Mazzarello, R.; Zhang, W. Chemical design principles for cache-type Sc-Sb-Te phase-change memory materials. *Chem. Mater.* **2019**, *31*, 4008–4015. [CrossRef]
42. Xu, Y.; Wang, X.; Zhang, W.; Schafer, L.; Reindl, J.; Vom Bruch, F.; Zhou, Y.; Evang, V.; Wang, J.J.; Deringer, V.L.; et al. Materials Screening for Disorder-Controlled Chalcogenide Crystals for Phase-Change Memory Applications. *Adv. Mater.* **2021**, *33*, 2006221. [CrossRef] [PubMed]
43. Zhou, Y.; Sun, L.; Zewdie, G.M.; Mazzarello, R.; Deringer, V.L.; Ma, E.; Zhang, W. Bonding similarities and differences between Y–Sb–Te and Sc–Sb–Te phase-change memory materials. *J. Mater. Chem. C* **2020**, *8*, 3646–3654. [CrossRef]
44. Sun, L.; Zhou, Y.; Wang, X.; Chen, Y.; Deringer, V.L.; Mazzarello, R.; Zhang, W. Ab initio molecular dynamics and materials design for embedded phase-change memory. *NPJ Comput. Mater.* **2021**, *7*, 29. [CrossRef]
45. Iwasaki, H.; Harigaya, M.; Nonoyama, O.; Kageyama, Y.; Takahashi, M.; Yamada, K.; Deguchi, H.; Ide, Y. Completely Erasable Phase Change Optical Disc II: Application of Ag-In-Sb-Te Mixed-Phase System for Rewritable Compact Disc Compatible with CD-Velocity and Double CD-Velocity. *Jpn. J. Appl. Phys.* **1993**, *32*, 5241–5247. [CrossRef]
46. Matsunaga, T.; Akola, J.; Kohara, S.; Honma, T.; Kobayashi, K.; Ikenaga, E.; Jones, R.O.; Yamada, N.; Takata, M.; Kojima, R. From local structure to nanosecond recrystallization dynamics in AgInSbTe phase-change materials. *Nat. Mater.* **2011**, *10*, 129–134. [CrossRef] [PubMed]
47. Salinga, M.; Carria, E.; Kaldenbach, A.; Bornhöfft, M.; Benke, J.; Mayer, J.; Wuttig, M. Measurement of crystal growth velocity in a melt-quenched phase-change material. *Nat. Commun.* **2013**, *4*, 2371. [CrossRef] [PubMed]
48. Zhang, W.; Ronneberger, I.; Zalden, P.; Xu, M.; Salinga, M.; Wuttig, M.; Mazzarello, R. How fragility makes phase-change data storage robust: Insights from ab initio simulations. *Sci. Rep.* **2014**, *4*, 6529. [CrossRef] [PubMed]
49. Morikawa, T.; Kurotsuchi, K.; Kinoshita, M.; Matsuzaki, N.; Matsui, Y.; Fujisaki, Y.; Hanzawa, S.; Kotabe, A.; Terao, M.; Moriya, H.; et al. Doped In-Ge-Te Phase Change Memory Featuring Stable Operation and Good Data Retention. In Proceedings of the 2007 IEEE International Electron Devices Meeting, Washington, DC, USA, 10–12 December 2007.
50. Spreafico, E.; Caravati, S.; Bernasconi, M. First-principles study of liquid and amorphous InGeTe2. *Phys. Rev. B* **2011**, *83*, 144205. [CrossRef]
51. Cecchini, R.; Martella, C.; Lamperti, A.; Brivio, S.; Rossi, F.; Lazzarini, L.; Varesi, E.; Longo, M. Fabrication of ordered Sb–Te and In–Ge–Te nanostructures by selective MOCVD. *J. Phys. D Appl. Phys.* **2020**, *53*, 144002. [CrossRef]
52. Song, K.-H.; Beak, S.-C.; Lee, H.-Y. Amorphous-to-crystalline phase transition of (InTe)x(GeTe) thin films. *J. Appl. Phys.* **2010**, *108*, 024506. [CrossRef]
53. Song, Z.; Wang, R.; Xue, Y.; Song, S. The "gene" of reversible phase transformation of phase change materials: Octahedral motif. *Nano Res.* **2021**. [CrossRef]

54. Wang, R.; Song, Z.; Song, W.; Xin, T.; Lv, S.; Song, S.; Liu, J. Phase-change memory based on matched Ge-Te, Sb-Te, and In-Te octahedrons: Improved electrical performances and robust thermal stability. *InfoMat* **2021**, *3*, 1008–1015. [CrossRef]
55. Maeda, Y.; Andoh, H.; Ikuta, I.; Minemura, H. Reversible phase-change optical data storage in InSbTe alloy films. *J. Appl. Phys.* **1988**, *64*, 1715. [CrossRef]
56. Miao, N.; Sa, B.; Zhou, J.; Sun, Z.; Blomqvist, A.; Ahuja, R. First-principles investigation on the phase stability and chemical bonding of mInSb·nInTe phase-change random alloys. *Solid State Commun.* **2010**, *150*, 1375–1377. [CrossRef]
57. Los, J.H.; Kühne, T.D.; Gabardi, S.; Bernasconi, M. First-principles study of the amorphous In3SbTe2 phase change compound. *Phys. Rev. B* **2013**, *88*, 174203. [CrossRef]
58. Deringer, V.L.; Zhang, W.; Rausch, P.; Mazzarello, R.; Dronskowski, R.; Wuttig, M. A chemical link between Ge–Sb–Te and In–Sb–Te phase-change materials. *J. Mater. Chem. C* **2015**, *3*, 9519–9523. [CrossRef]
59. Heßler, A.; Wahl, S.; Leuteritz, T.; Antonopoulos, A.; Stergianou, C.; Schön, C.-F.; Naumann, L.; Eicker, N.; Lewin, M.; Maß, T.W.W.; et al. In3SbTe2 as a programmable nanophotonics material platform for the infrared. *Nat. Commun.* **2021**, *12*, 924. [CrossRef]
60. Park, S.J.; Park, S.-J.; Park, D.; An, M.; Cho, M.-H.; Kim, J.; Na, H.; Park, S.h.; Sohn, H. High-mobility property of crystallized In-Te chalcogenide materials. *Electron. Mater. Lett.* **2012**, *8*, 175–178. [CrossRef]
61. Zhang, W.; Deringer, V.L.; Dronskowski, R.; Mazzarello, R.; Ma, E.; Wuttig, M. Density functional theory guided advances in phase-change materials and memories. *MRS Bull.* **2015**, *40*, 856–865. [CrossRef]
62. Kühne, T.; Krack, M.; Mohamed, F.; Parrinello, M. Efficient and Accurate Car-Parrinello-like Approach to Born-Oppenheimer Molecular Dynamics. *Phys. Rev. Lett.* **2007**, *98*, 066401. [CrossRef]
63. Hutter, J.; Iannuzzi, M.; Schiffmann, F.; VandeVondele, J. cp2k:atomistic simulations of condensed matter systems. *WIREs Comput. Mol. Sci.* **2014**, *4*, 15–25. [CrossRef]
64. Perdew, J.P.; Burke, K.; Ernzerhof, M. Generalized gradient approximation made simple. *Phys. Rev. Lett.* **1996**, *77*, 3865–3868. [CrossRef]
65. Goedecker, S.; Teter, M.; Hutter, J. Separable dual-space Gaussian pseudopotentials. *Phys. Rev. B* **1996**, *54*, 1703. [CrossRef]
66. Kresse, G.; Hafner, J. Ab initio molecular dynamics for liquid metals. *Phys. Rev. B* **1993**, *47*, 558–561. [CrossRef] [PubMed]
67. Kresse, G.; Joubert, D. From ultrasoft pseudopotentials to the projector augmented-wave method. *Phys. Rev. B* **1999**, *59*, 1758. [CrossRef]
68. Deringer, V.L.; Tchougreeff, A.L.; Dronskowski, R. Crystal orbital Hamilton population (COHP) analysis as projected from plane-wave basis sets. *J. Phys. Chem. A* **2011**, *115*, 5461–5466. [CrossRef]
69. Maintz, S.; Deringer, V.L.; Tchougréeff, A.L.; Dronskowski, R. LOBSTER: A tool to extract chemical bonding from plane-wave based DFT. *J. Comput. Chem.* **2016**, *37*, 1030–1035. [CrossRef] [PubMed]
70. Nelson, R.; Ertural, C.; George, J.; Deringer, V.L.; Hautier, G.; Dronskowski, R. LOBSTER: Local orbital projections, atomic charges, and chemical-bonding analysis from projector-augmented-wave-based density-functional theory. *J. Comput. Chem.* **2020**, *41*, 1931–1940. [CrossRef] [PubMed]
71. Henkelman, G.; Arnaldsson, A.; Jónsson, H. A fast and robust algorithm for Bader decomposition of charge density. *Comput. Mater. Sci.* **2006**, *36*, 354–360. [CrossRef]
72. Caravati, S.; Bernasconi, M.; Parrinello, M. First principles study of the optical contrast in phase change materials. *J. Phys. Condens. Matter* **2010**, *22*, 315801. [CrossRef]
73. Wełnic, W.; Wuttig, M.; Botti, S.; Reining, L. Local atomic order and optical properties in amorphous and laser-crystallized GeTe. *Comptes Rendus Phys.* **2009**, *10*, 514–527. [CrossRef]
74. Ahmed, S.; Wang, X.; Zhou, Y.; Sun, L.; Mazzarello, R.; Zhang, W. Unraveling the optical contrast in Sb2Te and AgInSbTe phase-change materials. *J. Phys. Photon.* **2021**, *3*, 03401. [CrossRef]
75. Wang, V.; Xu, N.; Liu, J.-C.; Tang, G.; Geng, W.-T. VASPKIT: A user-friendly interface facilitating high-throughput computing and analysis using VASP code. *Comput. Phys. Commun.* **2021**, *267*, 108033. [CrossRef]
76. Grimme, S.; Antony, J.; Ehrlich, S.; Krieg, H. A consistent and accurate ab initio parametrization of density functional dispersion correction (DFT-D) for the 94 elements H-Pu. *J. Chem. Phys.* **2010**, *132*, 154104. [CrossRef]
77. Grimme, S.; Ehrlich, S.; Goerigk, L. Effect of the damping function in dispersion corrected density functional theory. *J. Comput. Chem.* **2011**, *32*, 1456–1465. [CrossRef]
78. Woolley, J.C. Solid Solution in the GeTe-InTe System. *J. Electrochem. Soc.* **1965**, *112*, 906. [CrossRef]
79. Deringer, V.L.; Zhang, W.; Lumeij, M.; Maintz, S.; Wuttig, M.; Mazzarello, R.; Dronskowski, R. Bonding nature of local structural motifs in amorphous GeTe. *Angew. Chem. Int. Ed.* **2014**, *53*, 10817–10820. [CrossRef] [PubMed]
80. Errington, J.R.; Debenedetti, P.G. Relationship between structural order and the anomalies of liquid water. *Nature* **2001**, *409*, 318–321. [CrossRef]
81. Caravati, S.; Bernasconi, M.; Kühne, T.D.; Krack, M.; Parrinello, M. Coexistence of tetrahedral- and octahedral-like sites in amorphous phase change materials. *Appl. Phys. Lett.* **2007**, *91*, 171906. [CrossRef]
82. Chen, Y.; Sun, L.; Zhou, Y.; Zewdie, G.M.; Deringer, V.L.; Mazzarello, R.; Zhang, W. Chemical understanding of resistance drift suppression in Ge-Sn-Te phase-change memory materials. *J. Mater. Chem. C* **2020**, *8*, 71–77. [CrossRef]
83. Sun, S.-Y.; Zhang, B.; Wang, X.-D.; Zhang, W. Density dependent local structures in InTe phase-change materials. *arXiv* **2021**, arXiv:2109.14203.

84. Mazzarello, R.; Caravati, S.; Angioletti-Uberti, S.; Bernasconi, M.; Parrinello, M. Signature of Tetrahedral Ge in the Raman Spectrum of Amorphous Phase-Change Materials. *Phys. Rev. Lett.* **2010**, *104*, 085503. [CrossRef]
85. Akola, J.; Jones, R.O. Structural phase transitions on the nanoscale: The crucial pattern in the phase-change materials Ge2Sb2Te5 and GeTe. *Phys. Rev. B* **2007**, *76*, 235201. [CrossRef]
86. Raty, J.-Y.; Zhang, W.; Luckas, J.; Chen, C.; Bichara, C.; Mazzarello, R.; Wuttig, M. Aging mechanisms of amorphous phase-change materials. *Nat. Commun.* **2015**, *6*, 7467. [CrossRef] [PubMed]
87. Siegrist, T.; Jost, P.; Volker, H.; Woda, M.; Merkelbach, P.; Schlockermann, C.; Wuttig, M. Disorder-induced localization in crystalline phase-change materials. *Nat. Mater.* **2011**, *10*, 202–208. [CrossRef]
88. Cobelli, M.; Galante, M.; Gabardi, S.; Sanvito, S.; Bernasconi, M. First-Principles Study of Electromigration in the Metallic Liquid State of GeTe and Sb2Te3 Phase-Change Compounds. *J. Phys. Chem. C* **2020**, *124*, 9599–9603. [CrossRef]
89. Xie, Y.; Kim, W.; Kim, Y.; Kim, S.; Gonsalves, J.; BrightSky, M.; Lam, C.; Zhu, Y.; Cha, J.J. Self-Healing of a Confined Phase Change Memory Device with a Metallic Surfactant Layer. *Adv. Mater.* **2018**, *30*, 1705587. [CrossRef] [PubMed]
90. Oh, S.H.; Baek, K.; Son, S.K.; Song, K.; Oh, J.W.; Jeon, S.-J.; Kim, W.; Yoo, J.H.; Lee, K.J. In situ TEM observation of void formation and migration in phase change memory devices with confined nanoscale Ge2Sb2Te5. *Nanoscale Adv.* **2020**, *2*, 3841–3848. [CrossRef]
91. Wuttig, M.; Bhaskaran, H.; Taubner, T. Phase-change materials for non-volatile photonic applications. *Nat. Photon.* **2017**, *11*, 465–476. [CrossRef]
92. Shportko, K.; Kremers, S.; Woda, M.; Lencer, D.; Robertson, J.; Wuttig, M. Resonant bonding in crystalline phase-change materials. *Nat. Mater.* **2008**, *7*, 653–658. [CrossRef]
93. Lencer, D.; Salinga, M.; Grabowski, B.; Hickel, T.; Neugebauer, J.; Wuttig, M. A map for phase-change materials. *Nat. Mater.* **2008**, *7*, 972–977. [CrossRef]
94. Huang, B.; Robertson, J. Bonding origin of optical contrast in phase-change memory materials. *Phys. Rev. B* **2010**, *81*, 081204. [CrossRef]
95. Wuttig, M.; Deringer, V.L.; Gonze, X.; Bichara, C.; Raty, J.Y. Incipient Metals: Functional Materials with a Unique Bonding Mechanism. *Adv. Mater.* **2018**, *30*, 1803777. [CrossRef] [PubMed]
96. Zhu, M.; Cojocaru-Mirédin, O.; Mio, A.M.; Keutgen, J.; Küpers, M.; Yu, Y.; Cho, J.-Y.; Dronskowski, R.; Wuttig, M. Unique Bond Breaking in Crystalline Phase Change Materials and the Quest for Metavalent Bonding. *Adv. Mater.* **2018**, *30*, 1706735. [CrossRef]
97. Raty, J.Y.; Schumacher, M.; Golub, P.; Deringer, V.L.; Gatti, C.; Wuttig, M. A Quantum-Mechanical Map for Bonding and Properties in Solids. *Adv. Mater.* **2019**, *31*, 1806280. [CrossRef] [PubMed]
98. Cheng, Y.; Cojocaru-Mirédin, O.; Keutgen, J.; Yu, Y.; Küpers, M.; Schumacher, M.; Golub, P.; Raty, J.-Y.; Dronskowski, R.; Wuttig, M. Understanding the Structure and Properties of Sesqui-Chalcogenides (i.e., V2VI3 or Pn2Ch3 (Pn = Pnictogen, Ch = Chalcogen) Compounds) from a Bonding Perspective. *Adv. Mater.* **2019**, *31*, 1904316. [CrossRef]
99. Kooi, B.J.; Wuttig, M. Chalcogenides by Design: Functionality through Metavalent Bonding and Confinement. *Adv. Mater.* **2020**, *32*, 1908302. [CrossRef] [PubMed]

Article

Crystallization and Electrical Properties of Ge-Rich GeSbTe Alloys

Stefano Cecchi [1,2,*], Iñaki Lopez Garcia [3], Antonio M. Mio [3], Eugenio Zallo [1,4], Omar Abou El Kheir [2], Raffaella Calarco [1,5], Marco Bernasconi [2], Giuseppe Nicotra [3] and Stefania M. S. Privitera [3]

1. Paul-Drude-Institut für Festkörperelektronik, Leibniz-Institut im Forschungsverbund Berlin e.V., Hausvogteiplatz 5–7, 10117 Berlin, Germany; eugenio.zallo@wsi.tum.de (E.Z.); raffaella.calarco@artov.imm.cnr.it (R.C.)
2. Department of Materials Science, University of Milano-Bicocca, via R. Cozzi 55, 20125 Milano, Italy; o.abouelkheir@campus.unimib.it (O.A.E.K.); marco.bernasconi@unimib.it (M.B.)
3. Institute for Microelectronic and Microsystems (IMM), National Research Council (CNR), Zona Industriale Ottava Strada 5, 95121 Catania, Italy; inakigarcia.lopez@imm.cnr.it (I.L.G.); antonio.mio@imm.cnr.it (A.M.M.); giuseppe.nicotra@imm.cnr.it (G.N.); stefania.privitera@imm.cnr.it (S.M.S.P.)
4. Walter Schottky Institut, Physik Department, Technische Universität München, Am Coulombwall 4, 85748 Garching, Germany
5. Institute for Microelectronic and Microsystems (IMM), National Research Council (CNR), Via del Fosso del Cavaliere 100, 00133 Roma, Italy
* Correspondence: stefano.cecchi@unimib.it

Abstract: Enrichment of GeSbTe alloys with germanium has been proposed as a valid approach to increase the crystallization temperature and therefore to address high-temperature applications of non-volatile phase change memories, such as embedded or automotive applications. However, the tendency of Ge-rich GeSbTe alloys to decompose with the segregation of pure Ge still calls for investigations on the basic mechanisms leading to element diffusion and compositional variations. With the purpose of identifying some possible routes to limit the Ge segregation, in this study, we investigate Ge-rich Sb_2Te_3 and Ge-rich $Ge_2Sb_2Te_5$ with low (<40 at %) or high (>40 at %) amounts of Ge. The formation of the crystalline phases has been followed as a function of annealing temperature by X-ray diffraction. The temperature dependence of electrical properties has been evaluated by in situ resistance measurements upon annealing up to 300 °C. The segregation and decomposition processes have been studied by scanning transmission electron microscopy (STEM) and discussed on the basis of density functional theory calculations. Among the studied compositions, Ge-rich $Ge_2Sb_2Te_5$ is found to be less prone to decompose with Ge segregation.

Keywords: Ge-rich alloys; crystallization temperature; segregation; electrical properties

1. Introduction

GeSbTe (GST) chalcogenide alloys exhibit phase change properties that make this class of materials suitable for application in non-volatile memory technology [1–4]. In memory devices, the amorphous and crystalline phases are employed as logic states due to the large resistance contrast. The switching from the crystalline to the amorphous phases is obtained by applying a short, high-energy electrical pulse that melts the material, which is followed by a rapid thermal quench. A longer pulse at lower intensity is employed to recrystallize the memory cell. Such a phase change memory (PCM) technology is well assessed and already in production in the 20 nm node, with high potential to dominate the storage class memory sector, combining persistence and speed in the same device [5]. Despite the increasing technological impact, there are still some challenges that need to be solved, such as the low data retention, which presently can be a limiting factor for automotive or embedded applications, and the resistance drift, mainly affecting the amorphous phase. The data retention is limited by the crystallization temperature. The most commonly adopted materials for PCM have compositions belonging to the pseudo binary GeTe-Sb$_2$Te$_3$ tie-line of the GST ternary phase diagram, since no phase separation occurs after phase switching. Along this line, the crystallization temperature rises from 140 to 190 °C, as the GeTe amount increases [6]; therefore, it is not high enough to guarantee operation at high temperature, such as those required for embedded memories and for automotive applications. With the purpose of identifying materials with better data retention, without compromising the switching speed, GST alloys with composition off the GeTe-Sb$_2$Te$_3$ tie-line have been extensively studied in the literature, and the crystallization temperature for different alloys has been measured [7–11]. A survey of different GST compositions is reported in Figure 1. Among them, Ge-rich alloys have been proposed as valid candidates to address high-temperature applications due to their high crystallization temperatures. In the following, the Ge content in the alloys refers to the atomic percentage (e.g., for Ge$_2$Sb$_2$Te$_5$ (GST225) at %(Ge) ≈ 22%).

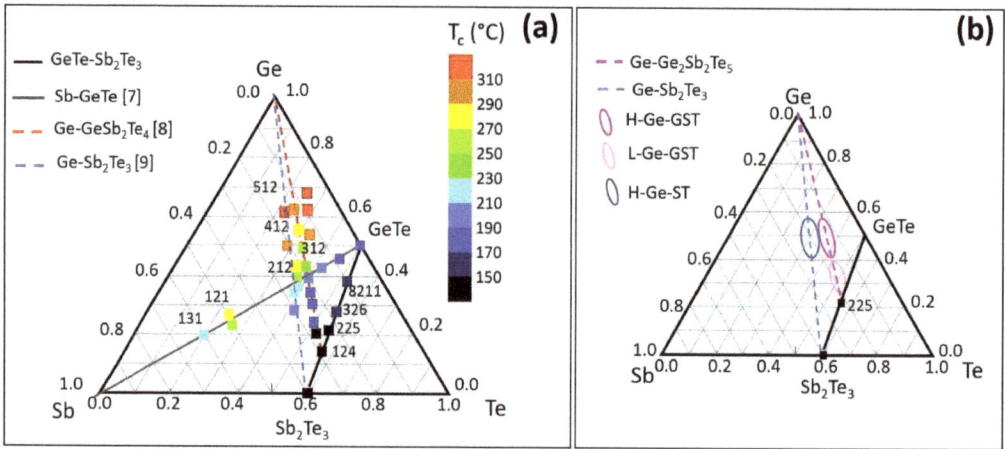

Figure 1. (a) Crystallization temperature of several compositions within the ternary GeSbTe phase diagram. The values have been taken from refs. [7–9]. (b) Compositions studied in this work.

In particular, Ge-rich Ge$_2$Sb$_1$Te$_2$ was proposed as "golden composition", since it showed a quite fast speed (about 80 ns) with a crystallization temperature of 250 °C [8]. Over the last years, several other compositions have been optimized in terms of cyclability, drift, and speed. However, the high crystallization temperature of Ge-rich alloys seems to be due to the slowing down of the crystallization kinetics produced by the mass transport involved in the phase separation process into crystalline Ge and GST alloys less rich in Ge [12,13]. Such a tendency to decompose still requires investigation on the basic mechanisms leading to compositional variations, since these may produce retention degradation upon cycling and even device failure. Recently, atomistic simulations applied to amorphous Ge-rich GST along the Ge-Ge$_1$Sb$_2$Te$_4$ line have revealed that these alloys tend to separate into Ge and Ge$_1$Sb$_2$Te$_4$ even in the amorphous phase when the Ge content is above 50% [14]. Thus, such a phase separation process is more likely to occur upon crystallization. The phase separation of Ge-Ge$_1$Sb$_2$Te$_4$ alloys during crystallization has been very recently studied by means of high-throughput density functional theory (DFT) calculations [15]. It has been shown that the Ge-rich Ge$_1$Sb$_2$Te$_4$ alloys become more prone to decompose into crystalline Ge and a GST alloy less rich in Ge, as Ge amount is above 50 at %. Therefore, it is crucial to understand to which extent the excess Ge segregates in Ge-rich GST alloys since, once impoverished in Ge, the remaining material may revert to stoichiometric phases that crystallize at lower temperatures. Moreover, the effect of the annealing thermal history on crystallization and segregation in Ge-rich GST has been shown by Di Biagio et al. [16]. The optimization of the microstructure of the crystallized films may influence also the switching properties.

In order to assess the relationship between amorphous stability and alloy composition in terms of structure and electrical properties, we investigate Ge-rich alloys with composition in the regions of the ternary phase diagram shown in Figure 1b. The electrical resistance as well as the crystallization process and the formed phases are investigated as a function of annealing temperature. The experimental observations have been corroborated by DFT calculations, which allow for the identification of the possible decomposition pathways during crystallization and therefore suggest some viable routes to limit the Ge segregation.

2. Materials and Methods

Amorphous chalcogenide Ge-rich GST films were deposited either on Si(111)-($\sqrt{3} \times \sqrt{3}$) R30°-Sb passivated surfaces or on SiO_2 substrates. The samples were fabricated by molecular beam epitaxy (MBE), and the substrate temperature was kept at room temperature. The film thickness was 30–60 nm, and the growth rate was ≈0.3–0.6 nm/min depending on film composition. As shown in Figure 1b, the studied compositions are Ge-rich Sb_2Te_3 with a high amount of Ge (>40 at %), which will be indicated in the following as H-Ge-ST; Ge-rich GST225, with a low (<40 at %) or high amount of Ge, which will be indicated in the following as L-Ge-GST and H-Ge-GST, respectively. For Ge-rich GST225 samples, the temperature of the Ge effusion cell was correspondingly set to achieve the two compositions, while all the other growth parameters were unchanged.

The films grown on Si(111) were capped in situ by a sputtered $ZnS:SiO_2$ layer and annealed ex situ by means of rapid thermal annealing (RTA). The annealing treatments were performed at increasing temperatures in the range of 140–330 °C for 30 min each under 1 bar nitrogen atmosphere. The RTA ramp-up time was set to 10 s for all temperatures. After each annealing step, the structural properties have been followed ex situ by X-ray diffraction (XRD) in order to study the crystallization and segregation processes.

For XRD measurements, a four-circle PANalytical X'Pert Pro Material Research Diffractometer system (Malvern Panalytical Ltd., Malvern, UK), equipped with a Ge (220) hybrid monochromator and Cu Kα1 X-ray radiation (λ = 1.540598 Å), has been employed. The curves were acquired in ω-2θ configuration along the symmetric rod of the Si(111) substrate.

The electrical properties have been investigated by in situ sheet resistance measurements of the samples grown on SiO_2 with a four-point probe, during annealing in air using a Temptronic ThermoChuck system (inTEST Thermal Solutions GmbH, Müllrose, Germany). The temperature dependence of the resistance has been evaluated in the range from 30 to 300 °C by employing a Agilent HP 4156B parameter analyser (Agilent Technologies, Inc., Santa Clara, CA, USA).

The crystalline structure and the element distribution after crystallization have been investigated by scanning transmission electron microscopy (STEM) and electron energy loss spectroscopy (EELS). The microscopical analyses have been performed by using a JEOL ARM200F (JEOL Ltd., Tokyo, Japan) Cs-corrected microscope, which is equipped with a cold-field emission gun and operating at 200 keV. Micrographs were acquired in Z-contrast mode by high-angle annular dark field (HAADF). A GIF Quantum ER system (Gatan AMETEK, Pleasanton, CA, USA) was used for EELS measurements. Both low- and core-loss EELS spectra were acquired with the Dual EELS tool through Gatan DigitalMicrograph software Version 3.4 (Gatan AMETEK, Pleasanton, CA, USA), in spectrum imaging (SI) mode.

Finally, we have computed the reaction free energy for the decomposition pathways of H-Ge-ST and H-Ge-GST alloys, by using the DFT data on the formation free energy of cubic alloys in the central part of the ternary Ge-Sb-Te phase diagram reported in our previous work, [15] which we refer to for all details.

3. Results and Discussion

Figure 2a,b shows the XRD ω-2θ scans acquired after different annealing steps at increasing temperature for H-Ge-ST and H-Ge-GST, respectively. The sharp peaks at ≈2, 4, and 6 Å^{-1} correspond to the 111, 222, and 333 Bragg reflections of the Si substrate. In both alloys, a GST phase with rocksalt structure is formed at low annealing temperatures (see peaks at ≈1.83, 3.66, 5.49 Å^{-1}) [17,18], whilst higher temperatures are required to observe the crystallization of segregated germanium (peaks at ≈1.93 and 5.79 Å^{-1} for 111 and 333 Bragg reflections, respectively). Although the two alloys have equal Ge content, the temperature at which crystallization occurs is different. In particular, in H-Ge-ST, the onset of crystallization of the GST rocksalt phase occurs at 140 °C, and the segregated Ge crystallizes at about 180 °C. For H-Ge-GST, crystallization shifts to higher temperatures, with the formation of rocksalt GST between 180 and 210 °C (crystallization onset still at 140 °C) and the Ge crystallization at 270 °C, respectively. In addition, we notice the intensity ratio between GST and Ge peaks in the two samples after crystallization, which qualitatively suggests their different tendency to decompose. The weak peak at ≈3.38 Å^{-1} that appears for H-Ge-ST at 210 °C may be attributed to segregated Sb, as confirmed by STEM and EELS analysis (see afterwards). Therefore, these data already suggest the higher stability of H-Ge-GST with respect to H-Ge-ST.

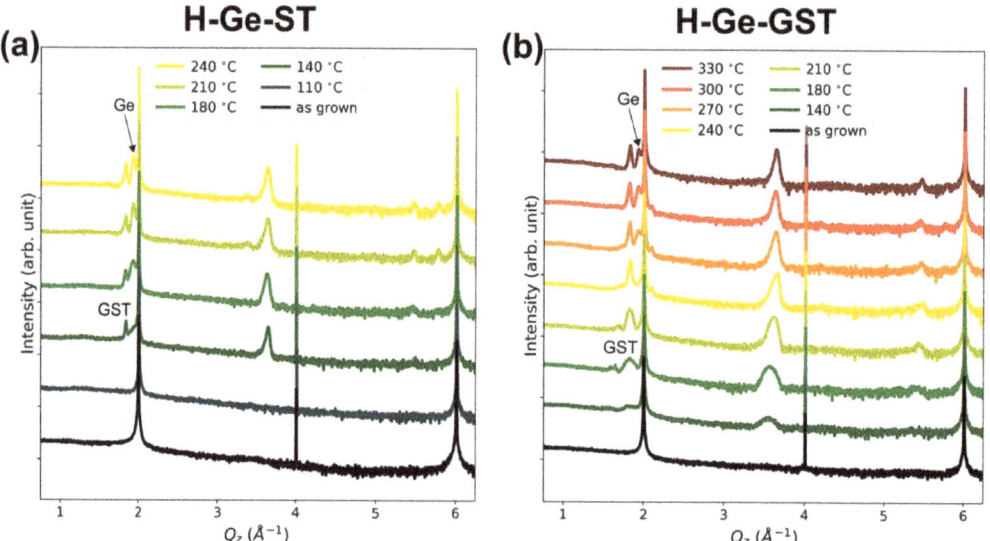

Figure 2. XRD ω-2θ scans as a function of annealing temperature for (**a**) H-Ge-ST (Ge-rich Sb$_2$Te$_3$), and (**b**) H-Ge-GST (Ge-rich Ge$_2$Sb$_2$Te$_5$).

XRD data of sample L-Ge-GST, upon annealing up to 300°C, indicate the presence of a single GST phase and no Ge segregation (see Figure S1 in the Supplementary Materials).

Figure 3a shows the resistivity versus annealing temperature as measured in situ after 3 min isothermal annealing for all the studied compositions. The resistivity of the as-grown amorphous phase increases as the Ge content increases, and the highest value is measured for H-Ge-GST. For the composition with lower Ge amount (L-Ge-GST), the temperature dependence of the resistivity of the amorphous phase corresponds to an activation energy of 0.45 eV, which is a value typical of amorphous Ge$_2$Sb$_2$Te$_5$ (GST225). At about 160 °C, the resistivity sharply decreases due to the film crystallization. Hence, consistently with previous studies, the crystallization temperature is higher with respect to a reference GST225 [9]. The resistance variation from amorphous to crystalline phase for L-Ge-GST is about four orders of magnitude.

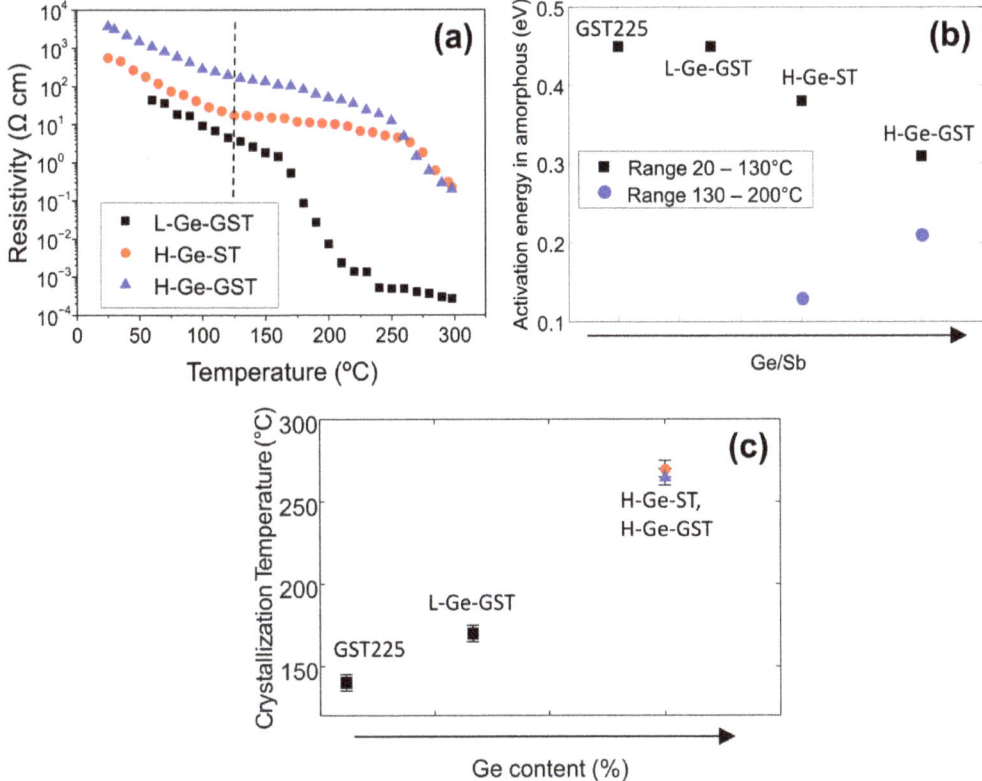

Figure 3. (a) Resistivity as a function of temperature for the studied compositions. (b) Activation energy for conductivity in the amorphous phase as a function of the ratio between Ge and Sb. (c) Crystallization temperature as obtained from the first derivative of the resistivity versus temperature as a function of Ge content.

For higher Ge content, the activation energy for the conductivity of the amorphous phase is slightly lower than that for GST225 and decreases as the Ge/Sb ratio decreases, as shown by black symbols in Figure 3b. Interestingly, at about 130 °C, the dependence of the resistivity versus temperature further reduces (see dashed line) and, as a consequence, the resistivity remains above 1 Ω cm up to 230 °C. The activation energies for amorphous conductivity obtained in the range 130–200 °C are shown in Figure 3b as blue symbols. In this case, the activation energy increases with the Ge/Sb ratio. These observations indicate that (i) up to 230 °C, the electrical conduction properties of H-Ge-GST and H-Ge-ST are mainly determined by the conductivity of the amorphous matrix; (ii) still, modifications of its composition are occurring, contextually to the nucleation of the GST crystallites, and such a process is more noticeable in H-Ge-ST. Finally, from the first derivative of the resistivity versus temperature, the effective crystallization temperature of the films has been evaluated as the temperature at which the derivative exhibits a minimum. It is worth noting that in the case of alloys in which different phases are formed upon annealing, as observed by XRD for H-Ge-GST and H-Ge-ST samples, the crystallization temperature obtained from the electrical measurements does not correspond to the crystallization of one single phase. Instead, it represents an effective value corresponding to the temperature at which a percolation path between regions with low resistance may be established. The results reported in Figure 3c clearly indicate that the effective crystallization temperature is dominated by the Ge amount. The two alloys with equal high Ge content, H-Ge-ST and H-Ge-GST, exhibit the same effective crystallization temperature within errors.

In the case of the alloys with higher Ge content, we also observe that the resistance contrast between the amorphous and the crystalline phases is about two orders of magnitude (see Figure 3a). This value, lower than that of GST225 and L-Ge-GST, is still enough to ensure distinguishing between two logic states [19]. However, we note that in these samples, the saturation of the low resistance value is not reached up to 300 °C. Due to limitations of the employed experimental apparatus, in situ annealing at temperatures above 300 °C could not be performed. In addition, TEM analysis of the sample H-Ge-ST annealed at 300 °C (see Figure S2 in the Supplementary Materials) shows that the film is still partially amorphous. Therefore, upon complete crystallization, the resistance contrast is expected to be larger than two orders of magnitude.

In order to get more insight in particular on the microstructure of the crystallized films, we have performed STEM and EELS analyses of the two alloys with higher Ge content. Figure 4a shows an STEM micrograph of the H-Ge-ST film on SiO_2 (uncapped) after annealing in the same setup used for electrical measurements at 180 °C for 30 min. The sample has been prepared aiming to investigate the modifications of the amorphous phase composition occurring at temperatures in the range 130–230 °C, as suggested by the electrical characterization. The upper part of the film shows several crystalline regions, whilst the bottom part is still mainly amorphous. The formation of a thin GeO layer at the film surface is also observed, suggesting that Ge oxidation may facilitate the crystallization of the underlying film, since it remains less rich in Ge. Figure 4b shows the elemental maps obtained by EELS spectra acquired in each pixel in SI mode. The elemental analysis shows the GeO topmost layer, appearing bright in the Ge map. In the same Ge map, the dark layer below the oxide is a crystalline region in which Ge is heavily depleted and Sb and Te correspondently exhibit higher concentrations. The crystalline regions, which form below the Ge-depleted region, contain less Ge atoms. On the contrary, higher Ge concentration is observed in the residual amorphous regions at the bottom of the film. This indicates that Ge atoms move (i) toward the surface, where they are oxidized, leaving regions depleted of Ge below the surface oxide; and (ii) toward the bottom of the film, enriching the amorphous fraction. Therefore, in the uncapped film, the Ge-depleted region may crystallize at lower temperature, favoring the further crystallization of GST. The reduction of the crystallization temperature in uncapped $Ge_2Sb_2Te_5$ films exposed to air, due to the selective oxidation of Ge, and to a minor extent of Sb, has been in fact extensively reported in the literature [20–26]. This behavior has been also recently confirmed in Ge-rich GST [27].

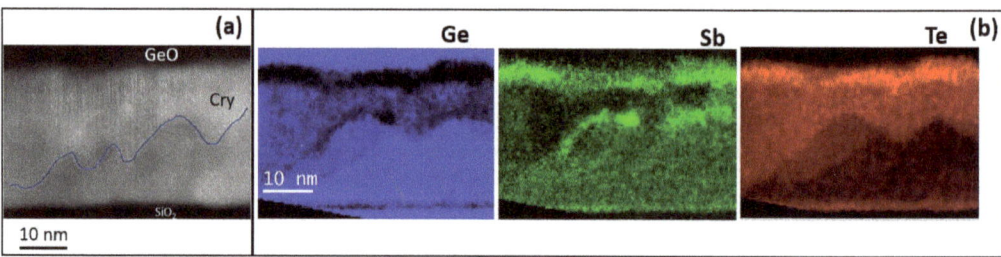

Figure 4. (**a**) STEM micrograph of the film H-Ge-ST after annealing in air at 180 °C. The crystallized (Cry) region is indicated in the figure. (**b**) Ge, Sb, and Te elemental maps obtained by EELS spectra acquired in SI mode. The intensity of the signal corresponding to the regions of Ge, Sb, and Te in the EELS spectra is reported in false colors.

This experiment provided evidence of the formation of less Ge-rich crystalline grains, in this case heterogeneously nucleating underneath the Ge-depleted region, which is compatible with the crystallization observed in the XRD data at temperatures below 180 °C.

Despite such evolution of the film microstructure, the results of the electrical measurements indicate that the electrical resistivity of the film remains very high up to 270 °C, while we record a change in the activation energy for the conductivity of the amorphous phase. This is because the crystalline grains, having lower Ge concentration, in order to grow expel the excess Ge, which therefore accumulates in the residual amorphous regions. Therefore, the amorphous fraction is expected to become richer and richer in Ge as the annealing temperature increases and the less-Ge rich crystalline grains grow. Still, as far as the crystalline grains are embedded into an amorphous matrix, and no percolation path is available, the electrical properties of the material are dominated by the part of the film which is still not crystallized. Ultimately, the change of activation energy highlighted in Figure 3a can be explained by considering the Ge enrichment of the amorphous matrix. Further details can be found in the Supplementary Materials: Figures S3 and S4.

Figure 5a shows an STEM micrograph of the same H-Ge-ST uncapped film on SiO$_2$ after annealing up to 300 °C for 30 min. The bottom part of the film is still partially amorphous (see also Figure S2). The higher contrast of some grains in Figure 5a is due to their alignment to the zone axis. By merging the elemental maps obtained by EELS, we extracted a ternary GST phase diagram, as shown in Figure 5b, and the corresponding stoichiometric map (Figure 5c). In the stoichiometric map of Figure 5c, regions with composition within a given range are plotted with the same color. The compositional ranges are shown in Figure 5b (dashed rectangles). The crystalline grains observed below the surface oxide have mainly a composition close to GeTe, with some Sb amount (yellow regions), whilst the bottom part of the film has a Ge-rich Sb$_2$Te$_3$ composition, as shown in green, with a higher Ge amount than that in the as-grown sample. Some regions, reported in red, have an Sb rich composition, which indicates that Sb tends to accumulate at the grain boundaries or in defective regions.

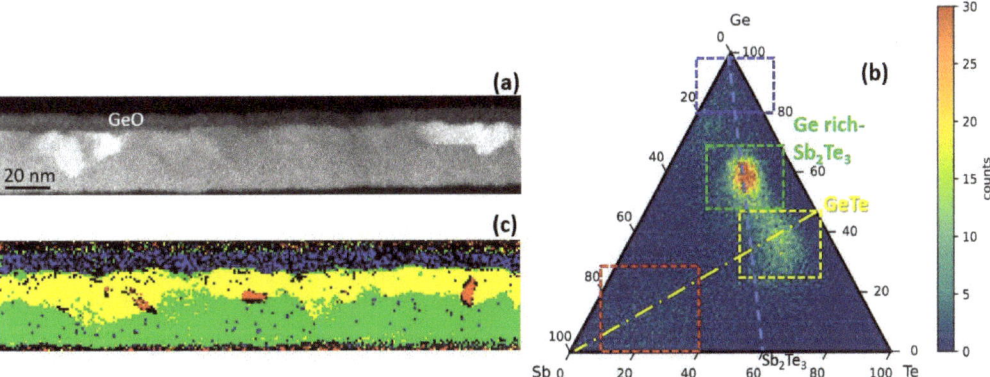

Figure 5. (a) STEM image of H-Ge-ST film on SiO$_2$ after annealing at 300 °C. (b) Ternary GeSbTe diagram built by merging the atomic maps distribution obtained by EELS spectra acquired in SI mode. (c) Stoichiometric map extracted from the ternary diagram. Ge-rich, Sb-rich, GeTe-rich, and Ge-rich Sb$_2$Te$_3$ regions are marked in blue, red, yellow, and green, respectively.

Figure 6a shows the STEM micrograph of the H-Ge-GST film deposited on Si(111) and annealed up to 330 °C for XRD characterization. For this sample, we recall that a ZnS:SiO$_2$ layer was deposited by in situ sputtering to prevent the oxidation during the annealing study. First, it is relevant to evaluate the overall crystallization status of the film. After the seven annealing steps from 140 to 330 °C, the material is completely crystalline, confirming our interpretation of the electrical measurements. Then, the ternary GST phase diagram extracted from the EELS elemental maps is reported in Figure 6b, and the distribution of the phases is shown in Figure 6c. The largest part of the film crystallizes preferentially with a composition of Ge$_{3+x}$Sb$_2$Te$_6$ (shown in light blue), which is close to the GeTe-Sb$_2$Te$_3$

pseudobinary line (reported in Figure 6b as black dashed line). Ge-rich regions (in blue) and with stoichiometry close to GeTe (in yellow) are also observed.

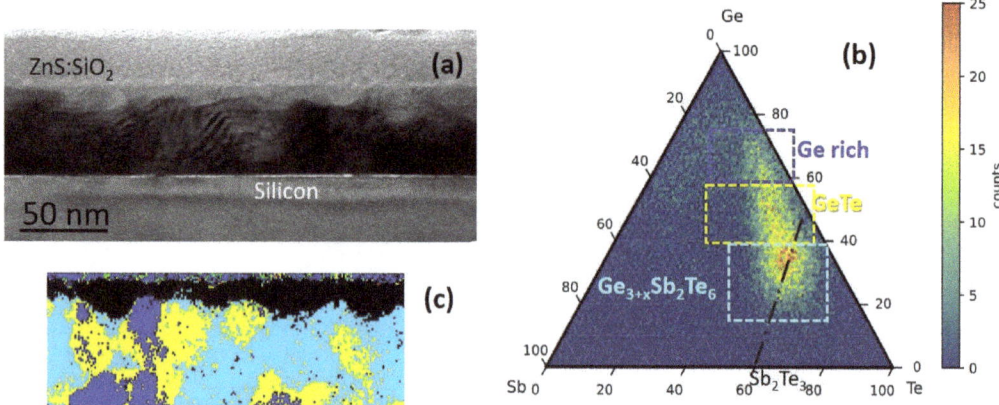

Figure 6. (a) STEM micrograph of H-Ge-GST on Si(111) after annealing at 330 °C. (b) Ternary GeSbTe diagram built by merging the atomic maps distribution obtained by EELS spectra acquired in SI mode. (c) Stoichiometric map extracted from the ternary diagram. Ge-rich, GeTe-rich, and $Ge_{3+x}Sb_2Te_6$ regions are marked in blue, yellow, and light blue, respectively.

Further TEM analyses, including selected area electron diffraction, are reported in Figures S5 and S6 in the Supplementary Materials.

The experimental findings can be better understood with the support of DFT calculations. In a previous work [15], we have computed the DFT formation free energy, with respect to the elements in their standard states, of cubic GST alloys in the central region of the ternary phase diagram. The free energy consisted of the total energy at zero temperature and the configurational free energy (at room temperature) due to disorder in the cubic sublattices. The calculated formation free energy is lower for alloys along the $GeTe$-Sb_2Te_3 tie-line and on the Sb-GeTe isoelectronic line [15].

This information can be used to compute the reaction free energy of the decomposition of a given cubic alloy with composition $Ge_xSb_yTe_z$ along all possible pathways given by

$$Ge_xSb_yTe_z \rightarrow a\ Ge_hSb_kTe_m + b\ Sb + c\ Ge + d\ Te + e\ Sb_2Te_3 + f\ GeTe \qquad (1)$$

Calculations on selected cases have shown that the vibrational contribution to the reaction free energy is negligible [28].

For each alloy, we can then construct a map of decomposition free energy, which highlights the more probable decomposition paths occurring during crystallization from the amorphous phase. In Ref. [15], this scheme was exemplified by studying the decomposition pathways of Ge-rich alloys on the Ge-$GeSb_2Te_4$ tie-line. Following the same methodology, we have computed the decomposition maps for the two alloys with high Ge amounts studied here.

Figure 7 shows the map of decomposition pathways during crystallization for H-Ge-ST (a) and H-Ge-GST (b). The decomposition paths have been calculated for the nominal stoichiometries of H-Ge-ST and H-Ge-GST samples. Each colored point for the generic alloy $Ge_hSb_kTe_m$ gives the value of the decomposition free energy for the formation of $Ge_hSb_kTe_m$. Negative reaction free energy indicates an exothermic reaction, while the green and blue regions correspond to the more favored decomposition pathways.

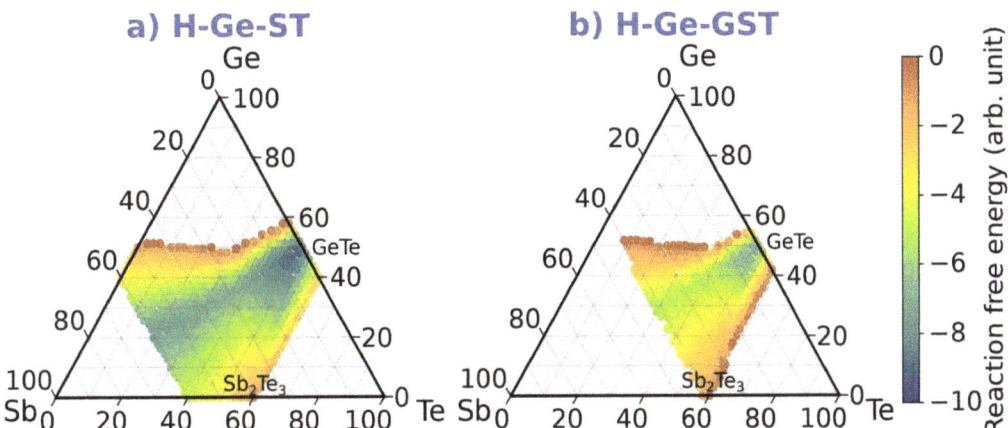

Figure 7. Color map of decomposition pathways during crystallization for (**a**) H-Ge-ST and (**b**) H-Ge-GST obtained from DFT data on the formation free energies in Ref. [15].

Only exothermic reactions forming $Ge_h Sb_k Te_m$ products that correspond at least to 33 at % in reaction (1) are reported in the map. Only under these conditions do we consider an alloy the main product of the decomposition pathway and show a point on the decomposition map.

We note that for H-Ge-ST, the reaction free energy is negative for a wider part of the phase diagram, indicating that several compositions may be formed. The most exothermic reactions lead to the formation of alloys on the Sb-GeTe and GeTe-Sb_2Te_3 lines, with a larger weight for alloys close to GeTe. Thus, we may expect the existence of different cubic alloys as the results of crystallization of amorphous Ge-rich Sb_2Te_3. In comparison, far fewer exothermic pathways are obtained for the decomposition of H-Ge-GST, suggesting that this alloy should have a lower tendency to decompose.

In our previous work [15], we introduced a measure of the decomposition propensity to quantify how much a given alloy is prone to decompose. This is obtained by counting the exothermic decomposition channels, weighted for their reaction free energy, including now all the possible products, even if present only in a small atomic fraction. The resulting map shows that the decomposition propensity is low along the GeTe-Sb_2Te_3 tie-line and also along the Sb-GeTe isoelectronic line [15]. In order to compare the theoretical results with the experimental data, and to evaluate possible strategies to tune the composition and obtain high crystallization temperature while minimizing the decomposition propensity, in Figure 8, we report as full squares the crystallization temperature of several compounds, as given in Figure 1. Circular symbols represent the calculated decomposition propensity. Blue points correspond to compositions with low propensity to decompose. L-Ge-GST (violet ellipse), that we found to exhibit no phase separation upon crystallization (see Figure S1), belongs to this group, together with other compositions that are characterized by crystallization temperature lower than 220 °C. Red circular points correspond to compositions with a high propensity to decompose. Alloys with higher crystallization temperature (>270 °C) are found within the red region, and therefore, they are expected to undergo massive phase separation upon crystallization. The composition H-Ge-ST (blue ellipse) is within a region with high propensity to decompose. The average fraction of Ge segregation for all alloys in the central part of the GeSbTe phase diagram was also computed in Ref. [15]. This map is shown in the inset on the left of Figure 8, which further highlights that a lower fraction of Ge segregates in the crystallization process of alloys on the Ge-$Ge_2Sb_2Te_5$ tie-line than for alloys on the Ge-Sb_2Te_3 line. It is worth noting that the high crystallization temperature of Ge-rich alloys has been ascribed to the phase separation process which, requiring the diffusion of the atomic species on a long length scale, slows down the crystallization kinetics [13]. However, such an explanation clearly limits the possibility to select an alloy with sufficiently high crystallization temperature that is able to homogeneously switch with no phase separation, which is desired to minimize the cell-to-cell variation and the resistance drift of the crystalline state but is jeopardized by the presence of nanocrystals with different compositions or by residual amorphous regions. To this end, a trade-off is represented by compositions lying at the edges between regions with high and low propensity to decompose. Among these, we find the composition Ge-rich $Ge_2Sb_1Te_2$ with Ge below 50 at %, which has been already identified in the past [8], and the H-Ge-GST (magenta ellipse) studied in this work. Both compositions are expected to segregate no more than 30% of the total amount of Ge. Nevertheless, the crystallization temperature is about 270 °C. Such a result suggests that the mass transport involved in the phase segregation may play a role in raising the crystallization temperature, but it cannot be the unique mechanism. Other factors should be taken into account, such as, for example, the dependence of the structural properties of the amorphous phase on composition and thermal history [16]. Indeed, it has been shown by previous analysis of DFT models that by increasing Ge [14,28] or Sb [29] content in respect to the $GeTe-Sb_2Te_3$ tie-line, the amorphous network becomes more dissimilar from the cubic crystal. This is because of the formation of more Ge-Ge bonds with tetrahedral geometry, at the expense of octahedral bonding, which characterizes the rocksalt phase. Moreover, according to molecular dynamics simulations of the crystallization process [4,30–32], the key process in the formation of cubic crystalline nuclei is the reorientation of the four-membered rings. By increasing the fraction of Ge and Sb, the number of four-membered rings decreases at the expense of longer rings. The presence of a larger amount of homopolar bonds and tetrahedra, promoted by the excess Ge, and the concomitant decrease in the number of four-membered rings might slow down the crystallization kinetics either by reducing the nucleation rate or the crystal growth velocity, which overall leads to an increase in the crystallization temperature. Further increase in the crystallization temperature without increasing the Ge segregation may also be achieved by different approaches, such as nitrogen or carbon doping.

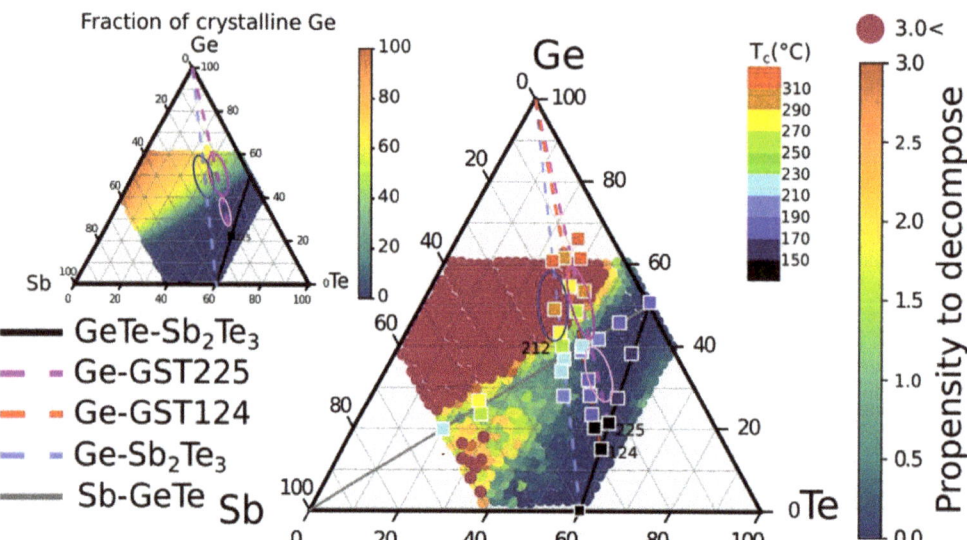

Figure 8. Crystallization temperature of several alloys studied in literature. Experimental data are plotted on the map obtained in Ref. [15] by DFT calculations showing the propensity to decompose (in eV/atom, see text). The inset on the left upper part shows the average crystalline Ge fraction upon crystallization, as obtained from DFT calculations in Ref. [15]. The compositions studied in this work, as already shown in Figure 1b, are indicated by ellipses.

4. Conclusions

In this paper, we have studied the stability and the electrical properties of Ge-rich GeSbTe alloys belonging to the Ge-Sb$_2$Te$_3$ and Ge-Ge$_2$Sb$_2$Te$_5$ lines. The Ge-rich GST225 film with Ge amount lower than 40 at % has a crystallization temperature of 180 °C and exhibits no phase separation. Both Ge-Sb$_2$Te$_3$ and Ge-Ge$_2$Sb$_2$Te$_5$ alloys with higher Ge amount (>40%) are characterized by the same higher crystallization temperature (270 °C) and by phase separation. However, depending on the composition, a different propensity to decompose is observed experimentally and evaluated by DFT calculations, with a lower fraction of segregated Ge for the composition H-Ge-GST. Therefore, this alloy is expected to have a better stability upon cycling in a phase change memory.

Supplementary Materials: The following are available online at https://www.mdpi.com/article/10.3390/nano12040631/s1, Figure S1: XRD ω-2θ scans obtained after annealing at 300 °C for H-Ge-GST (solid line) and L-Ge-GST (dotted line), Figure S2: High-resolution TEM image of H-Ge-ST after annealing at 300 °C for 30 min, Figure S3: Calculated resistivity of GST225 and Sb$_2$Te$_3$ according to the effective medium approximation, assuming the presence of some amount of excess Ge, Figure S4: Calculated resistivity of Sb$_2$Te$_3$ as a function of temperature, according to the effective medium approximation, assuming the presence of different amounts of excess Ge, Figure S5: (a) Bright field TEM and (b) SAED of the sample H-Ge-ST on SiO$_2$ after annealing in air at 180 °C. The SAED is obtained from a selected area of about 100 nm of diameter, Figure S6: (a) Bright field TEM and (b) SAED of the sample H-Ge-GST on Si(111) after annealing at 330 °C. The SAED is obtained from a selected area of about 100 nm of diameter. Reflections marked in yellow correspond to the Si substrate along the [110] zone axis. The other reflections (in red) are compatible with Sb$_2$Te$_3$ and trigonal (GeTe)$_n$(Sb$_2$Te$_3$).

Author Contributions: Conceptualization, S.C., S.M.S.P. and M.B.; Methodology, S.C., I.L.G., E.Z., A.M.M. and O.A.E.K.; Investigation, S.C., I.L.G., A.M.M., S.M.S.P., O.A.E.K. and M.B.; Resources, G.N.; Data Curation, S.C., A.M.M., M.B. and S.M.S.P.; Writing—Original Draft Preparation, S.M.S.P.; Writing—Review and Editing, S.C. and M.B.; Funding Acquisition, R.C. All authors have read and agreed to the published version of the manuscript.

Funding: This work has received funding from the European Union's Horizon 2020 research and innovation program under grant agreement No 824957—BEFOREHAND Project (Boosting Performance of Phase Change Devices by Hetero- and Nano-structure Material Design) and under grant agreement No 823717—ESTEEM3 Project.

Institutional Review Board Statement: Not applicable.

Informed Consent Statement: Not applicable.

Data Availability Statement: The data that support the findings of this study are available from the corresponding author upon reasonable request.

Acknowledgments: K. Morgenroth is acknowledged for performing the annealing and characterization measurements of L-Ge-GST and H-Ge-GST. The authors thank S. Behnke and C. Stemmler for technical support at the molecular beam epitaxy system, S. Pannitteri for technical support in TEM samples preparation, M. Borghi and A. Redaelli for fruitful discussion.

Conflicts of Interest: The authors declare no conflict of interest.

References

1. Wuttig, M.; Yamada, N. Phase-change materials for rewriteable data storage. *Nat. Mater.* **2007**, *6*, 824–832. [CrossRef]
2. Raoux, S.; Wełnic, W.; Ielmini, D. Phase change materials and their application to nonvolatile memories. *Chem. Rev.* **2010**, *110*, 240–267. [CrossRef]
3. Burr, G.W.; Breitwisch, M.J.; Franceschini, M.; Garetto, D.; Gopalakrishnan, K.; Jackson, B.; Kurdi, B.; Lam, C.; Lastras, L.A.; Padilla, A.; et al. Phase change memory technology. *J. Vac. Sci. Technol.* **2010**, *28*, 223. [CrossRef]
4. Zhang, W.; Mazzarello, R.; Wuttig, M.; Ma, E. Designing crystallization in phase-change materials for universal memory and neuro-inspired computing. *Nat. Rev. Mater.* **2019**, *4*, 150–168. [CrossRef]
5. Freitas, F.R.; Wilcke, W.W. Storage-class memory: The next storage system technology. *IBM Res. J. Dev.* **2008**, *52*, 439. [CrossRef]

6. Yamada, N.; Ohno, E.; Nishiuchi, K.; Akahira, N.; Takao, M. Rapid-phase transitions of GeTe-Sb$_2$Te$_3$ pseudobinary amorphous thin films for an optical disk memory. *J. Appl. Phys.* **1991**, *69*, 2849–2856. [CrossRef]
7. Cheng, H.-Y.; Raoux, S.; Wuttig, M.; Munoz, B.; Jordan-Sweet, L.J. The Crystallization Behavior of Ge$_1$Sb$_x$Te$_1$ Phase-Change Materials. In Proceedings of the 2010 MRS Spring Meeting, San Francisco, CA, USA, 5–9 April 2010; p. H 6.5.
8. Cheng, H.Y.; Hsu, T.H.; Raoux, S.; Wu, J.; Du, P.Y.; Breitwisch, M.; Zhu, Y.; Lai, E.K.; Joseph, E.; Mittal, S.; et al. A high performance phase change memory with fast switching speed and high temperature retention by engineering the Ge$_x$Sb$_y$Te$_z$ phase change material. In Proceedings of the 2011 International Electron Devices Meeting, Washington, DC, USA, 5–7 December 2011; pp. 3.4.1–3.4.4. [CrossRef]
9. Zuliani, P.; Palumbo, E.; Borghi, M.; Dalla Libera, G.; Annunziata, R. Engineering of chalcogenide materials for embedded applications of Phase Change Memory. *Solid. Stat. Electron.* **2015**, *111*, 27–31. [CrossRef]
10. Sousa, V.; Navarro, G.; Castellani, N.; Coué, M.; Cueto, O.; Sabbione, C.; Noé, P.; Perniola, L.; Blonkowski, S.; Zuliani, P.; et al. *Symposium on VLSI Technology (Digest of Technical Papers)*; IEEE: Piscataway, NJ, USA, 2015; Volume 7.4, p. T98.
11. Kiouseloglou, A.; Navarro, G.; Sousa, V.; Persico, A.; Roule, A.; Cabrini, A.; Torelli, G.; Maitrejean, S.; Reimbold, G.; De Salvo, B.; et al. A novel programming technique to boost low-resistance state performance in Ge-rich GST phase change memory. *IEEE Trans. Electron. Devices* **2014**, *61*, 1246. [CrossRef]
12. Navarro, G.; Coué, M.; Kiouseloglou, A.; Noé, P.; Fillot, F.; Delaye, V.; Persico, A.; Roule, A.; Bernard, M.; Sabbione, C. Trade-off between SET and data retention performance thanks to innovative materials for phase-change memory. In Proceedings of the 2013 IEEE International Electron Devices Meeting, Washington, DC, USA, 9–11 December 2013; pp. 21.5.1–21.5.4. [CrossRef]
13. Agati, M.; Vallet, M.; Joulié, S.; Benoit, D.; Claverie, A. Chemical phase segregation during the crystallization of Ge-rich GeSbTe alloys. *J. Mater. Chem.* **2019**, *7*, 8720–8729. [CrossRef]
14. Sun, L.; Zhou, Y.-X.; Wang, X.-D.; Chen, Y.-H.; Deringer, V.L.; Mazzarello, R.; Zhang, W. Ab initio molecular dynamics and materials design for embedded phase-change memory. *NPJ Comput. Mater.* **2021**, *7*, 29. [CrossRef]
15. Abou El Kheir, O.; Bernasconi, M. High-throughput calculations on the decomposition reactions of off-stoichiometry GeSbTe alloys for embedded memories. *Nanomaterials* **2021**, *11*, 2382. [CrossRef]
16. Di Biagio, F.; Cecchi, S.; Arciprete, F.; Calarco, R. Crystallization Study of Ge-Rich (GeTe)$_m$(Sb$_2$Te$_3$)$_n$ Using Two-Step Annealing Process. *Phys. Stat. Sol. Rapid Res. Lett.* **2019**, *13*, 1800632. [CrossRef]
17. Bragaglia, V.; Arciprete, F.; Zhang, W.; Mio, A.M.; Zallo, E.; Perumal, K.; Giussani, A.; Cecchi, S.; Boschker, J.E.; Riechert, H. Metal—Insulator Transition Driven by Vacancy Ordering in GeSbTe Phase Change Materials. *Sci. Rep.* **2016**, *6*, 23843. [CrossRef]
18. Cecchi, S.; Zallo, E.; Momand, J.; Wang, R.; Kooi, B.J.; Verheijen, M.A.; Calarco, R. Improved structural and electrical properties in native Sb$_2$Te$_3$/Ge$_x$Sb$_2$Te$_{3+x}$ van der Waals superlattices due to intermixing mitigation. *APL Mater.* **2017**, *5*, 026107. [CrossRef]
19. Zuliani, P.; Varesi, E.; Palumbo, E.; Borghi, M.; Tortorelli, I.; Erbetta, D.; Dalla Libera, G.; Pessina, N.; Gandolfo, A.; Prelini, C.; et al. Overcoming Temperature Limitations in Phase Change Memories With Optimized Ge$_x$Sb$_y$Te$_z$. *IEEE Trans. Electron. Devices* **2013**, *60*, 4020. [CrossRef]
20. Noé, P.; Vallée, C.; Hippert, F.; Fillot, F.; Raty, J.-Y. Phase-change materials for non-volatile memory devices: From technological challenges to materials science issues. *Semicond. Sci. Technol.* **2018**, *33*, 013002. [CrossRef]
21. Noé, P.; Sabbione, C.; Bernier, N.; Castellani, N.; Fillot, F.; Hippert, F. Impact of interfaces on scenario of crystallization of phase change materials. *Acta Mater.* **2016**, *110*, 142. [CrossRef]
22. Kooi, B.J.; Groot, W.M.; De Hosson, J.T. In situ transmission electron microscopy study of the crystallization of Ge$_2$Sb$_2$Te$_5$. *J. Appl. Phys.* **2004**, *95*, 924. [CrossRef]
23. Berthier, R.; Bernier, N.; Cooper, D.; Sabbione, C.; Hippert, F.; Noé, P. In situ observation of the impact of surface oxidation on the crystallization mechanism of GeTe phase-change thin films by scanning transmission electron microscopy. *J. Appl. Phys.* **2017**, *122*, 115304. [CrossRef]
24. Bragaglia, V.; Mio, A.M.; Calarco, R. Thermal annealing studies of GeTe-Sb$_2$Te$_3$ alloys with multiple interfaces. *AIP Adv.* **2017**, *7*, 085113. [CrossRef]
25. Gourvest, E.; Pelissier, B.; Vallee, C.; Roule, A.; Lhostis, S.; Maitrejean, S. Impact of Oxidation on Ge$_2$Sb$_2$Te$_5$ and GeTe Phase-Change Properties. *J. Electrochem. Soc.* **2012**, *159*, H373.
26. Kim, Y.; Park, A.S.; J.Baeck, H.; M.Noh, K.; Jeong, K. Phase separation of a Ge$_2$Sb$_2$Te$_5$ alloy in the transition from an amorphous structure to crystalline structures. *J. Vac. Sci. Technol.* **2006**, *24*, 929. [CrossRef]
27. Agati, M.; Gay, C.; Benoit, D.; Claverie, A. Effects of surface oxidation on the crystallization characteristics of Ge-rich Ge-Sb-Te alloys thin films. *Appl. Surface Sci.* **2020**, *518*, 146227. [CrossRef]
28. Abou El Kheir, O.; Dragoni, D.; Bernasconi, M. Density functional simulations of decomposition pathways of Ge-rich GeSbTe alloys for phase change memories. *Phys. Rev. Mat.* **2021**, *5*, 095004. [CrossRef]
29. Gabardi, S.; Caravati, S.; Bernasconi, M.; Parrinello, M. Density functional simulations of Sb-rich GeSbTe phase change alloys. Condens. *J. Phys. Condens. Matter.* **2012**, *24*, 385803. [CrossRef]
30. Kalikka, J.; Akola, J.; Jones, R.O. Crystallization processes in the phase change material Ge$_2$Sb$_2$Te$_5$: Unbiased density functional/molecular dynamics simulations. *Phys. Rev. B* **2016**, *94*, 134105. [CrossRef]

31. Sosso, G.C.; Miceli, G.; Caravati, S.; Giberti, F.; Behler, J.; Bernasconi, M. Fast Crystallization of the Phase Change Compound GeTe by Large-Scale Molecular Dynamics Simulations. *J. Phys. Chem. Lett.* **2013**, *4*, 4241–4246. [CrossRef]
32. Hegedüs, J.; Elliott, S.R. Microscopic origin of the fast crystallization ability of Ge–Sb–Te phase-change memory materials. *Nat. Mater.* **2008**, *7*, 399–405. [CrossRef]

Article

Growth, Electronic and Electrical Characterization of Ge-Rich Ge–Sb–Te Alloy

Adriano Díaz Fattorini [1,2], Caroline Chèze [2], Iñaki López García [3], Christian Petrucci [1], Marco Bertelli [1], Flavia Righi Riva [2], Simone Prili [2], Stefania M. S. Privitera [3], Marzia Buscema [3], Antonella Sciuto [3], Salvatore Di Franco [3], Giuseppe D'Arrigo [3], Massimo Longo [1], Sara De Simone [1], Valentina Mussi [1], Ernesto Placidi [2,4], Marie-Claire Cyrille [5], Nguyet-Phuong Tran [5], Raffaella Calarco [1,*] and Fabrizio Arciprete [2]

1. Istituto per la Microelettronica e Microsistemi (IMM), Consiglio Nazionale delle Ricerche (CNR), Via del Fosso del Cavaliere 100, 00133 Rome, Italy; adriano.diazfattorini@artov.imm.cnr.it (A.D.F.); chridapa@gmail.com (C.P.); marco.bertelli@artov.imm.cnr.it (M.B.); massimo.longo@artov.imm.cnr.it (M.L.); sara.desimone@artov.inm.cnr.it (S.D.S.); valentina.mussi@artov.imm.cnr.it (V.M.)
2. Dipartimento di Fisica, Università di Roma "Tor Vergata", Via della Ricerca Scientifica 1, 00133 Rome, Italy; cheze_caroline@yahoo.fr (C.C.); friva@roma2.infn.it (F.R.R.); simone.prili@roma2.infn.it (S.P.); ernesto.placidi@uniroma1.it (E.P.); fabrizio.arciprete@roma2.infn.it (F.A.)
3. Istituto per la Microelettronica e Microsistemi (IMM), Consiglio Nazionale delle Ricerche (CNR), Zona Industriale Ottava Strada 5, 95121 Catania, Italy; inaki.lopez@outlook.com (I.L.G.); stefania.privitera@imm.cnr.it (S.M.S.P.); marzia.buscema@st.com (M.B.); antonella.sciuto@imm.cnr.it (A.S.); salvatore.difranco@imm.cnr.it (S.D.F.); giuseppe.darrigo@imm.cnr.it (G.D.)
4. Department of Physics, Sapienza University of Rome, P.le Aldo Moro 5, 00185 Rome, Italy
5. Leti, CEA, University Grenoble Alpes, 38000 Grenoble, France; marie-claire.cyrille@cea.fr (M.-C.C.); nguyet-phuong.tran@cea.fr (N.-P.T.)
* Correspondence: raffaella.calarco@artov.imm.cnr.it

Abstract: In this study, we deposit a Ge-rich Ge–Sb–Te alloy by physical vapor deposition (PVD) in the amorphous phase on silicon substrates. We study in-situ, by X-ray and ultraviolet photoemission spectroscopies (XPS and UPS), the electronic properties and carefully ascertain the alloy composition to be GST 29 20 28. Subsequently, Raman spectroscopy is employed to corroborate the results from the photoemission study. X-ray diffraction is used upon annealing to study the crystallization of such an alloy and identify the effects of phase separation and segregation of crystalline Ge with the formation of grains along the [111] direction, as expected for such Ge-rich Ge–Sb–Te alloys. In addition, we report on the electrical characterization of single memory cells containing the Ge-rich Ge–Sb–Te alloy, including I-V characteristic curves, programming curves, and SET and RESET operation performance, as well as upon annealing temperature. A fair alignment of the electrical parameters with the current state-of-the-art of conventional $(GeTe)_n$-$(Sb_2Te_3)_m$ alloys, deposited by PVD, is found, but with enhanced thermal stability, which allows for data retention up to 230 °C.

Keywords: PCM; Ge-rich GST alloys; Raman; electronic properties

Citation: Díaz Fattorini, A.; Chèze, C.; López García, I.; Petrucci, C.; Bertelli, M.; Righi Riva, F.; Prili, S.; Privitera, S.M.S.; Buscema, M.; Sciuto, A.; et al. Growth, Electronic and Electrical Characterization of Ge-Rich Ge–Sb–Te Alloy. *Nanomaterials* **2022**, *12*, 1340. https://doi.org/10.3390/nano12081340

Academic Editor: Julian Maria Gonzalez Estevez

Received: 18 March 2022
Accepted: 11 April 2022
Published: 13 April 2022

Publisher's Note: MDPI stays neutral with regard to jurisdictional claims in published maps and institutional affiliations.

Copyright: © 2022 by the authors. Licensee MDPI, Basel, Switzerland. This article is an open access article distributed under the terms and conditions of the Creative Commons Attribution (CC BY) license (https://creativecommons.org/licenses/by/4.0/).

1. Introduction

In recent years, the interest of the industrial and research sectors has been increasingly directed towards the so-called "Internet of Things" (IoT) [1]. The IoT is a new way of conceiving the world around us, and it aims to connect people with each other, people with things, and things together. One of the fields in which IoT is most widespread is certainly the automotive one, the purpose of which is the development of car automation, in order to improve safety, emissions, and production costs. The idea is to be able to connect cars with each other, as well as with road devices, buildings, cyclists, and even pedestrians, all this to make driving safer. To reach this level of intelligent driving, it is necessary to have devices capable of acquiring, storing, and sending data, while being cheap, fast in reading and writing, non-volatile, and resistant over the years. This is one of the biggest challenges in this sector. Phase change materials (PCM) are the materials of choice for such applications [2]. The peculiarities of PCM-based memory devices are a low-current for the phase transition, good reading and writing speed, and low production costs. Among the materials most used in such devices, there is $(GeTe)_n$-$(Sb_2Te_3)_m$ (GST), a ternary alloy consisting of two binary compounds, GeTe and Sb_2Te_3, especially in the composition $Ge_2Sb_2Te_5$ (GST225) [3]. The use of GST for automotive is a great challenge, above all because, in some parts of the vehicles, temperatures up to 300 °C can be reached (in the vicinity of the exhaust gases even 800 °C) also in conditions of stationary operations. Therefore, the devices must be able to work at high temperatures (at least 160 °C) to keep the data stable for at least ten years [4]. One of the disadvantages of using GST225 is its crystallization temperature (about 150 °C), which does not favor good thermal stability and, therefore, data retention for automotive applications. When considering materials for automotive applications, it must also be borne in mind that the memories undergo, for short times, end-of-process treatments, which occur at around 400 °C. In recent years, several attempts to improve the crystallization temperature (T_x) were made by doping the GST sample with C [5,6], O [7,8], N [9,10], or Ge itself, the so-called Ge-rich GST [11,12]. The T_x of Ge-rich GST was demonstrated to increase almost linearly with Ge content and has proven data retention after soldering reflow in industrial integrated device [11]. Nevertheless, as very recently pointed out by Redaelli et al. [13], although Ge-rich GST solved the issue of high temperature retention requests, a large degree of material management, in terms of material instabilities, remains. Finally, it is worth mentioning that In–Sb–Te and In–Ge–Te phase-change alloys are also of interest, since, for high indium contents, they exhibit higher thermal stability of the amorphous phase, with respect to the Ge–Sb–Te alloys. The crystallization temperature is increased up to 290 °C, in the case of $In_3Sb_1Te_2$ [14], and 276 °C, in the case of doped In–Ge–Te, for which 10-years retention at temperatures higher than 150 °C has been found [15]. Reports on In-based alloys, both in the form of planar PCM devices [16] and as single and core-shell nanowires [17,18], have been published. Therefore, it is important to investigate the possibility of using appropriate material alloys and combinations to improve on material management issues.

In this work, we present an extensive characterization of Ge-rich GST, of composition GST 29 20 28, with the assessment of its quality. Electronic and compositional properties are investigated by photoemission characterization, vibrational by Raman spectroscopy, and structural by X-ray diffraction. The electrical characterization confirmed the high working temperature in single memory devices, obtained using a GST225 buffer layer underneath the Ge-rich GST. The novelty of the present work is the usage of such a buffer layer, which improves the electrical bottom contact and guarantees the thermal stability typical of Ge-rich GST.

2. Experimental

2.1. Sample Growth

The PCM films were deposited on Si(001)/SiO$_2$ substrates via physical vapor deposition (PVD) in a custom-made, ultra-high vacuum (UHV) chamber system, equipped with Te, Sb, and Ge (Alfa Aisar, Haverhill, MA, USA) loaded in Knudsen cells (Dr. Eberl MBE-Komponenten GmbH, Weil der Stadt Germany) for thermal evaporation. The pressure during the growth reached high 10^{-9} Torr. In order to grow completely amorphous films, the substrate temperature was kept at nominal room temperature (RT). However, due to radiative heating from the cell crucibles, the substrate temperature was actually in the range 85–90 °C. Thus, some faint crystallization might be obtained. The fluxes ratios were kept to Ge:Sb:Te = 1:1:3 and Ge:Sb:Te = 2.5:1:1.8, respectively, for achieving nominal GST225 and Ge-rich GST alloy compositions. The growth time was 17 and 30 min for a nominal deposited thickness of 24 nm for both GST films. The same growth conditions were used for the realization of a GST225 (24 nm)/Ge-rich GST (24 nm) double-layer heterostructure on a single-cell vehicle for electrical testing.

2.2. Photoemission Characterization

Photoemission spectra were acquired using a UHV chamber, dedicated to X-ray and ultraviolet photoelectron spectroscopy (XPS and UPS) and connected in UHV to the PVD growth chamber. XPS spectroscopy was performed using an Omicron DAR 400 Al/Mg Kα non monochromatized X-ray source (Taunusstein, Germany). To collect and analyze photoelectrons, a 100 mm hemispherical VG-CLAM2 electron spectrometer (Uckfield, UK) with a single channeltron and 4 mm entrance slit was used. XPS experiments were performed with a pass energy E_{pass} = 20 eV, which means an instrumental resolution of ΔE = 0.4 eV, to be convoluted with the natural linewidth of the sources: ΔE_{Al} = 0.84 eV and ΔE_{Mg} = 0.68 eV. In order to have the best possible resolution, all the XPS spectra were collected by using the Mg Kα radiation, except the one around the Ge 2p core levels, which would be superimposed on the secondary electrons background tail, for which, we used the Al Kα radiation. UPS (Uckfield, UK) experiments were performed with a pass energy E_{pass} = 1 eV, which means an instrumental resolution of ΔE = 0.02 eV to be convoluted with the natural linewidth of the sources: ΔE_{HeI} = 0.003 eV. All XPS spectra were analyzed and fitted by means of the KolXPD software (version 1.8.0) and libraries (http://kolxpd.com accessed on 8 August 2018) using Voigt peaks and a Shirley background.

2.3. X-ray Diffraction

X-ray diffraction (XRD) measurements were performed ex-situ after the deposition of a 10 nm thick protective W capping layer and successive annealing for 15 min at increasing temperatures, from 30 to 400 °C. XRD was carried out using a Bruker (Billerica, MA, USA) D8 Discover diffractometer, equipped with a Cu X-ray source and Anton Paar DHS1100 (Graz, Austria) dome-type heating stage for temperature measurements in N_2 atmosphere. Grazing incidence diffraction (GID) scans, after final cooling at RT, were acquired under an angle of 0.8°.

2.4. Raman

A Raman imaging DXR2xi microscope from Thermo Scientific (Waltham, MA, USA) was used to carry out the Raman measurements. The operation of the device is based on data acquisition, carried out in "backscattering" geometry, using a green laser beam with wavelength λ = 532 nm, as well as acquiring the data with a 50× magnification objective. For all the measurements carried out, the same experimental conditions were used, namely exposure time (t = 0.1 s), number of acquisitions (n = 200), and laser power on the sample (P = 4 mW).

2.5. Memory Device Realization

A classical mushroom architecture was selected to verify the memory functionalities. A cylindrical heater in titanium nitride was used as the back contact, and a tungsten layer, defined by lift-off procedure, was used as the top contact. The chalcogenide layer was laterally defined through plasma etching in an inductively coupled plasma system, with a gas mixture of CHF_3:Ar (25:30 sccm) having a radio frequency power of 200 W and bias of 75 V. The hard mask was fabricated in hydrogen silsesquioxane resist (Dow Corning® (Midland, MI, USA) XR-1541 e-beam resist), patterned by electron beam lithography in an E-Line Raith apparatus (Dortmund, Germany). The same e-beam equipment was used to inspect the memory devices used in the modality scanning electron microscope.

2.6. Electrical Measurements

The electrical characterization of PCM devices has been performed by using a Keysight (Santa Rosa, CA, USA) MSO64B oscilloscope and Keysight 81110A pulse pattern generator. The first RESET operation was achieved by applying a train of 300 ns pulses with increasing voltage (staircase up). Each programming pulse has been followed by a reading pulse at lower voltage (0.2 V). After reaching a maximum voltage, determined as the voltage at which the device switch is observed, the voltage pulse is decreases (staircase down).

3. Results and Discussion

In Figure 1, we show a schematic of the samples and experimental processes presented in the following of the present work. Photoemission spectroscopy was used in situ for the identification of elements near the surface of the Ge-rich GST, obtaining information on the local chemical environments and valence band electronic structure. The stoichiometry and concentration of elements were also obtained by a quantitative analysis of the XPS data. This spectroscopy is also a valuable tool for providing information on phase separation. In Figure 1a, the schematic for the XPS and UPS experiment is provided.

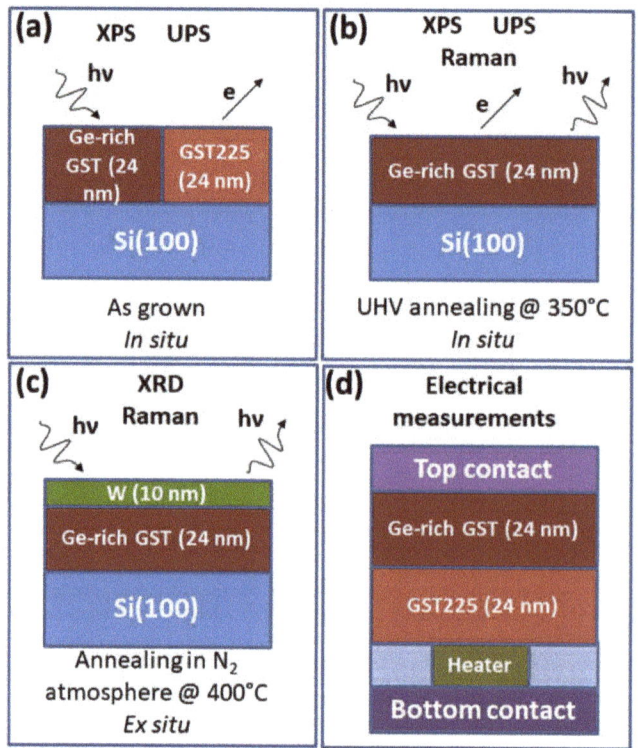

Figure 1. Schematics of samples and experiments carried out within this work: (**a**) XPS and UPS experiment on as grown sample; (**b**) XPS, UPS, and Raman measurements on as grown sample, after UHV annealing at 350 °C; (**c**) XRD and Raman investigation after crystallization, by means of annealing at 400 °C in N$_2$ atmosphere; (**d**) electrical measurements of the GST225/Ge-rich GST memory devices.

In Figure 2, we report the comparison between the shallow core levels of (a) Ge-rich GST and (b) GST225 sample, collected after annealing for 30 min in UHV at 350 °C (see Figure 1b).

From the analysis of the shallow core levels, we calculated the composition of the alloys obtaining $Ge_{18}Sb_{26}Te_{56}$, compatible with GST 14 20 43, as well as $Ge_{38}Sb_{26}Te_{36}$, compatible with GST 29 20 28, for GST225 and Ge-rich GST, respectively. A comparison between the binding energy (BE) of the core levels, as determined by the fitting of the XPS spectra, revealed that the Sb 4d doublet of Ge-rich GST shifts to BE lower than that of GST225; conversely, Te 4d and Ge 3d core levels both move at higher BE, compared to GST225. The general trend of the chemical shifts, observed in the XPS spectra, reflects the differences in the composition between the two alloys, as already observed, in the case of

Ge-rich GST-based heterostructure, as reported in this special issue [19]. A chemical shift of the core levels, after changing the GST alloy composition, is an initial state effect, which is related to a change in chemical bonding [20–22]. After the formation of a ternary Ge–Sb–Te alloy, the BE of Te 4d core levels should be in between the values expected for metallic Te and Sb_2Te_3. The more the alloy is Ge-rich, the more Te 4d core levels are shifted towards higher BEs. The same holds for Ge 3d core levels. In the case of Sb, 4d levels shift towards lower BEs are expected. The BE shifts observed in our samples agree with this general trend, if we consider the expected absolute BE positions for pure Te 4d, Sb 4d, and Ge 3d core levels (40.6 eV, 32.1 eV, and 29.3 eV, respectively).

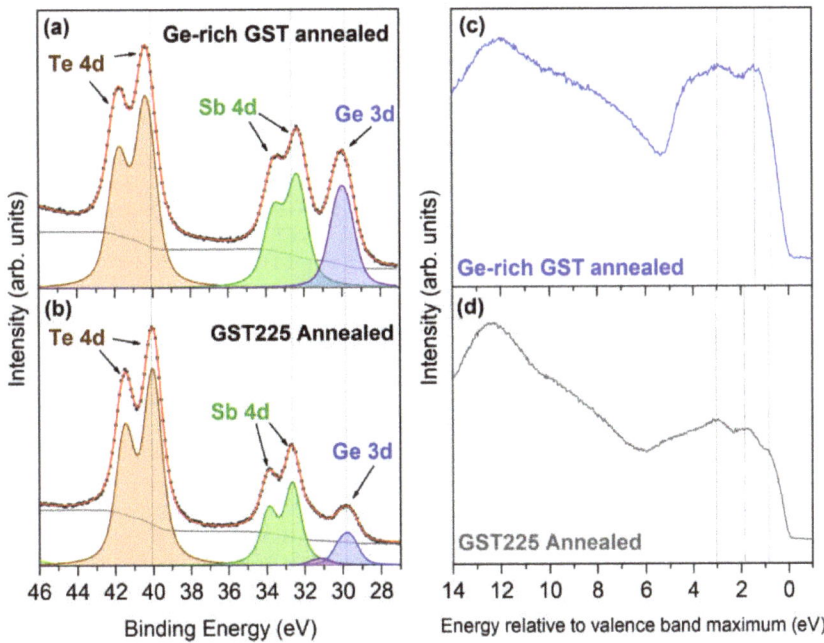

Figure 2. XPS spectra of (**a**) Ge-rich GST and (**b**) GST225 after annealing. Experimental data are presented using black dots, while red and gray lines correspond to the total data fit and Shirley background. Contributions from plasmon excitations, $K\alpha_3$ and $K\alpha_4$ lines are omitted for clarity. UPS spectra of (**c**) Ge-rich GST and (**d**) GST225 after annealing.

Furthermore, from the inspection of the UPS spectra in Figure 2c,d a clear difference in the valence band (VB) line shape can be seen. We could identify the shallow core levels Te 5s at about 12 eV, as well as Sb 5s and Ge 4s, located in the region between 10 and 6 eV, respectively. Therefore, it is evident that s electrons are quite deep and do not participate in the valence band structure. The proper valence band is observed between 6 and 0 eV and is a mix of Te 5p, Sb 5p, and Ge 4p states. In particular, the GST225 VB is characterized by three main features, at 0.85, 1.87, and 3.03 eV, as well as a broad structure extending from 1 to 6 eV, compatible with the VB of crystalline GST225 [21,23–25] and observed for a GST225/GST-based heterostructure [19]. In addition, the Ge-rich GST spectrum shows a shoulder at 0.72 eV, a clear peak at 1.41 eV, and broad double feature in the range 3–4 eV, typical of amorphous Ge-rich GST alloys. A comparison with literature calculation is pretty well in agreement [26]. The UPS analysis suggests that the Ge-rich GST sample is predominantly amorphous, even after annealing in UHV, as already observed by the XPS data analysis. This finding is evidence that the sample is amorphous, even after annealing in UHV. It is important to note that crystallization in UHV is not straightforward and strongly depends on the GST composition. In a recent paper, we proved efficient annealing in UHV of GST225 films. Indeed, the T_x for Ge-rich GST is expected to be higher than for GST225, so that an annealing temperature of 350 °C was used in the attempt to crystallize Ge-rich GST; such treatment might have induced desorption, instead of crystallization.

Theoretical calculations on the stability of Ge-rich GST have been carried out by O. Abou El Kheir et al. using density functional theory [26,27], showing that amorphous Ge-rich GST alloy is metastable, while its crystalline form is expected to be extremely unstable and decomposes in different alloys. Among others, the formation of GST323 is very likely. The high T_x is due to the segregation of crystalline Ge upon crystallization of the amorphous phase [28]. The mass transport involved in this phase separation slows down the crystallization kinetics, which leads to a higher T_x.

In Figure 3a, we present the Raman characterization of the Ge-rich GST sample, after removing it from the UHV (where it was subjected to annealing), compared with a reference amorphous GST225 alloy. Two peaks, at 106 and 154 cm^{-1}, and a broad peak, at 60–80 cm^{-1}, are visible. The latter peak is positioned at the same shift of the $A_{1g}(1)$ mode of Sb_2Te_3 [29]. However, unlike GST225, Ge-rich GST displays a sharp edge at 106 cm^{-1}, followed by a decreasing slope at the same position of the E_g mode of crystalline GST225, thus, possibly indicating a very faint crystallization of the Ge-rich GST, due to the annealing procedure. The peak at 123.6 cm^{-1}, which is present in GST225, is less prominent for the Ge-rich GST alloy [30]. Furthermore, in Ge-rich GST, a broad feature at 215 cm^{-1}, typical of the stretching of the Ge–Ge bond in the tetrahedral configuration is present [31]. We also note the absence of bands assigned to the TO and LO vibrational modes of amorphous Ge, usually present from 190 to 300 cm^{-1} [32], suggesting that the UHV annealings were not sufficient to enable the segregation of Ge in the amorphous phase. Furthermore, the absence of any additional peak at around 300 cm^{-1}, to be ascribed to pure crystalline Ge, is the indication that the excess of Ge was not crystallized at the temperature used for the annealing performed in the UHV system. Therefore, even from Raman investigations, we do not have an indication of a full crystallization of both Ge-rich GST and Ge [33,34]. Our interpretation is compatible with a recent discussion on the Ge-rich GST crystallization by L. Prazakova et al. [35]. They studied, by Raman spectroscopy and XRD techniques, the crystallization mechanism of Ge-rich GST alloys with a large Ge enrichment (15–55 at. %), where the structural evolution process in temperature begins with the Ge–Te bond rearrangement around stable SbTe structural units, independently of the Ge content in the alloy.

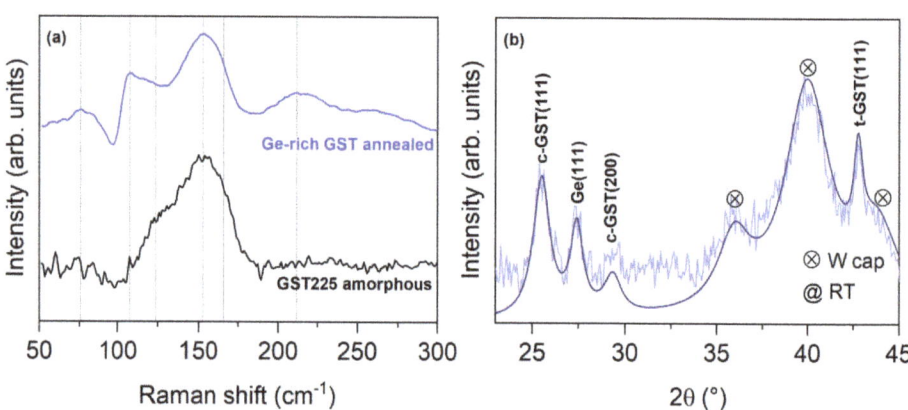

Figure 3. (a) Raman spectra of the Ge-rich GST and GST225 alloy. Dashed vertical lines are a guide to the eye for the main features; (b) 2θ GID curve of the W/Ge-rich GST/SiO$_2$/Si(001)stack, acquired at RT, after crystallization up to 400 °C. Light blue: experimental data; dark blue: fitted curve, using Lorentzian peaks.

From XPS measurements, we calculated the actual composition of the deposited Ge-rich GST to be GST 29 20 28, as obtained from flux ratios Ge:Sb:Te = 2.5:1:1.8. The percentage of the deposited Ge is 37.6%; therefore, it is important to investigate the segregation of Ge, as well as of the Ge-rich GST decomposition, as recently reported by Cecchi et al. [36]. Furthermore, we learned from Raman measurements that the excess of amorphous Ge is not segregating in the as-grown samples, thus ensuring that the deposition of amorphous GST 29 20 28 layers in devices will be homogeneous.

To crystallize the sample and observe the expected phase separation [13], we resorted to annealing in the N_2 atmosphere from RT to 400 °C, at steps of 50 °C, waiting 15 min at each step (see Figure 1c). The 2θ° GID curve shown in Figure 3b is acquired after cooling down to room temperature. Several diffraction peaks can be observed. The three peaks corresponding to the W capping layer are marked by crossed circles. The peaks at θ = 25.52° and 29.36° are identified as (111) and (200) Bragg reflections of the GST in the cubic crystalline phase, while the peak θ = 42.77° is ascribed to the trigonal crystalline phase. Interestingly, the peak at θ = 27.44 is the contribution of the (111) reflection of Ge, showing the Ge segregation upon annealing [37]. An additional consideration is on the amount of crystallized Ge that results rather small, as inferred from the overall intensity of the XRD and, more evidently, Raman peaks. Again, in this case, our observations are compatible with the results from L. Prazakova et al. [35], who reported that the crystallization of the cubic GST phase is followed by the heterogeneous nucleation and consequent growth of the Ge crystalline phase. This sequential crystallizations have been well-observed in layers with low Ge content, since they take place at different temperatures. On the contrary, the two processes appear simultaneously at high temperatures in the layers with high Ge content, resulting from delayed GST phase crystallization and fast Ge crystal growth.

Our annealing and XRD data provided the information (data not shown here) that the crystallization occurred completely at 350 °C; hence, it is important to know the annealing temperature of memory devices prior to the forming step (see below). The observation that both trigonal and cubic lattices coexist in the crystalline phase after annealing, thus demonstrating that the annealing of Ge-rich GST is not sufficient to transform the whole volume into the trigonal phase, is rather interesting. Such a finding will be the subject of further investigation on several Ge-rich GST samples with different compositions.

To further evaluate the material quality of the Ge-rich GST material for electrical memory applications, a duplicate sample was prepared in a device structure. Figure 4 shows a secondary electron microscopy (SEM) image of the top contact of the fabricated devices.

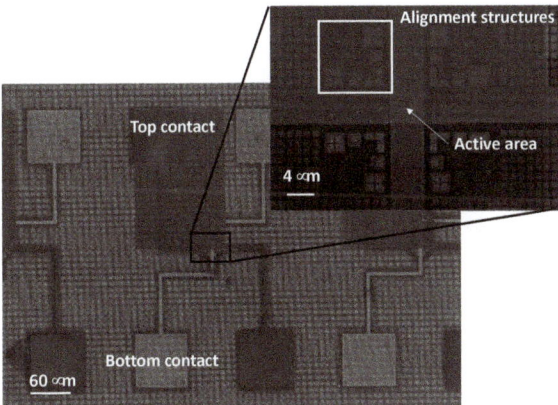

Figure 4. SEM microscope image of the memory devices. Bottom and top contacts are clearly visible. The blow-up shows a detail with evident alignment markers of the area of crossing of the bottom and top contacts, where the active material deposited (GST225/Ge-rich GST) is present, together with the nanometric TiN heater structure.

The Ge-rich GST sample was fabricated on a GST225 amorphous buffer of 24 nm (see Figure 1d). Such a buffer was deposited to obtain a better contact with the bottom of the device; as for Ge-rich compositions, we expect to have a higher resistivity. We also performed measurements of devices with GST225 as the active material, for which the succesful annealing temperature in the nitrogen atmosphere was 180 °C for 30 min. For GST225/Ge-rich GST, the annealing temperature of 180 °C was not sufficient to produce crystallization. Therefore, according to our structural characterization, the devices were annealed at 350 °C, in order to obtain the full crystallization of the phase change layer. Figure 3 shows the main results of GST225/Ge-rich GST devices with a heater diameter of 100 nm.

Figure 5a,b shows the current–voltage characteristics obtained during the staircase, up and down, and resistance measured, as a function of programming current, respectively. The I-V curve displays a threshold voltage V_{TH} = 2.3 V. The resistance is initially high and, after the "forming" process, it decreases during the staircase down. The forming process is suggested for Ge-rich GST materials. After forming, the device is programmed from SET to RESET (as shown in Figure 5c), from RESET to SET, and, again, to RESET, in order to determine the programming window, as shown in Figure 5d.

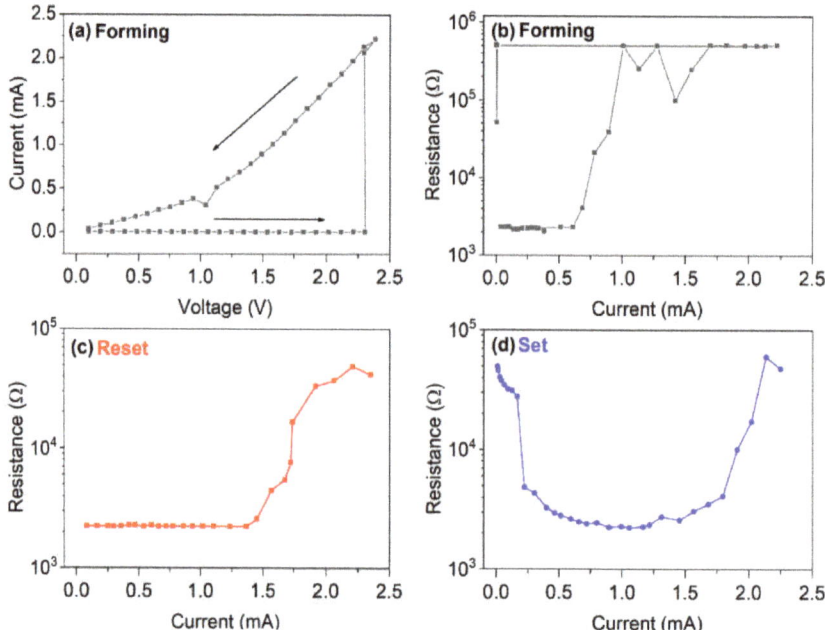

Figure 5. Current–voltage characteristics (**a**), resistance measured after each pulse versus programming current (**b**), as acquired during forming of the GST225/Ge-rich GST devices. Programming window of RESET (**c**) and SET (**d**) processes.

Figure 6 shows, as main reliability figures of merit, the SET and RESET resistance measured upon cycling (a) and resistance drift, as a function of time, in (b). The programming window resulted in more than one order of magnitude. The best measured endurance value was 2×10^4 cycles. The resistance of the devices in the RESET or SET state after 1h annealing, as a function of temperature, is reported in Figure 6c. The devices are able to retain information up to 230 °C. The presence of GST225 does not adversely affect the thermal stability. Typically, the GST225 device is not able to retain the information up to 230 °C for 1 h. Upon forming, it is reasonable to assume that some interdiffusion between the two layers, with migration of Ge atoms toward the GST225, might occur with the formation of an alloy with intermediate properties. In fact, the crystallization temperature is about 350 °C, while retention is only 230 °C. Figure 6d reports the scheme of staircase, up and down, used to measure the devices and position at which the read operation was performed.

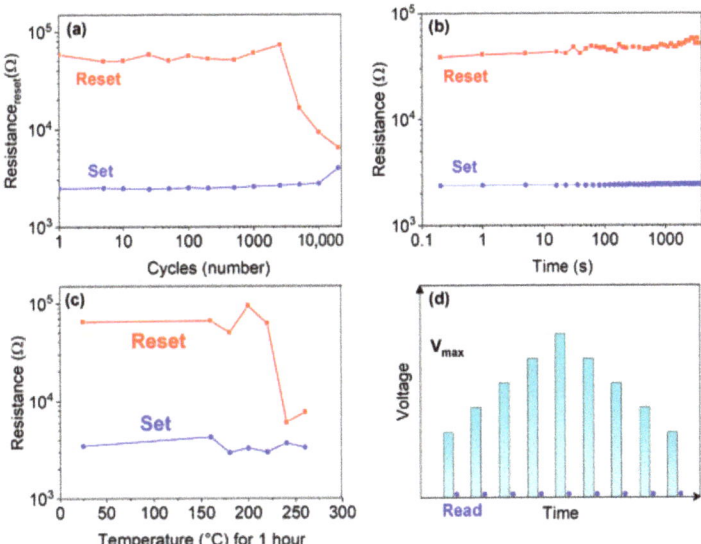

Figure 6. Resistance of the GST225/Ge-rich GST devices: upon cycling (**a**); versus time (**b**); after 1 h annealing at different temperatures (**c**); scheme of measurements staircase, up and down (**d**).

Phase change memory devices manufactured with a single Ge-rich GST layer have been studied in literature [11,13,38,39]. Despite the high tendency of Ge to segregate, alloys with Ge contents of 20–45% or more in excess, compared to the GST225 stoichiometry, are preferentially adopted, since the high Ge content may guarantee higher crystallization temperatures [11] However, the improved thermal stability is usually associated with some undesirable effects. One of these is the increase of the resistivity of the crystalline phase, increasing as the Ge content increases, as well as the corresponding increase of the drift coefficient, even in the SET state [39]. According to Ref. [39], single-cell memories with GST with an excess Ge of 35% (namely a composition close to the one adopted in this paper) exhibit a drift coefficient ν of 0.02 in the low resistance state, higher than that of GST225 (ν = 0.005). The resistance drift of both the SET and RESET states in the devices here presented is comparable to that typically measured in GST225. It is important to note that, although the thickness of the GST225 is comparable to that of the Ge-rich GST, and, even assuming some intermixing, the larger overall amount of Ge is dominating the device behavior, in terms of thermal stability. Therefore, data indicate that the presence of a GST225 buffer layer can successfully mitigate the increased resistance drift, while assuring data retention up to 230 °C (Figure 6c).

4. Conclusions

In this work, we present a full characterization of a PVD grown amorphous Ge-rich GST alloy. We evaluated the composition of the alloy to be $Ge_{36}Sb_{26}Te_{38}$, compatible with GST 29 20 28. Although annealing in UHV was conducted up to 350 °C, no crystallization of the film occurred. UPS measurements revealed a good agreement between the experimental spectra and theoretical calculations from the literature [26].

A segregation of crystalline Ge was observed, as well as the coexistence of the trigonal and cubic lattices in the crystalline phase, after annealing of Ge-rich GST.

The investigated Ge-rich GST alloy was integrated into a PCM single-cell testing vehicle, allowing for electrical characterizations; a V_{TH} = 2.3 V and programming window of about two orders of magnitude can be obtained. The resistance drift and endurance (2×10^4 cycles) turned out to be comparable to that of GST225. The electrical characterizations showed that the Ge-rich GST alloy works in a similar fashion as conventional GST225-based devices, except for the fact that the former retain the information up to 230 °C. The data retention of 230 °C, after 1 h annealing, is markedly higher than that of GST225 (about 160 °C), but lower than expected, according to the crystallization temperature (350 °C). We suggest that interdiffusion occurs during the first melting (the forming process), with migration of Ge atoms toward the GST225 and the formation of an alloy with intermediate properties.

Author Contributions: Conceptualization, R.C. and F.A.; methodology, R.C. and F.A.; investigation, A.D.F., C.C., I.L.G., C.P., M.B. (Marco Bertelli), F.R.R., S.P., M.B. (Marzia Buscema), A.S., S.D.F., G.D., M.L., S.D.S., V.M., E.P., M.-C.C. and N.-P.T.; resources, R.C., V.M., G.D., S.M.S.P., M.-C.C. and F.A.; data curation, R.C., F.A., A.D.F., F.R.R., S.P., I.L.G. and S.M.S.P.; writing—original draft preparation, R.C.; writing—review and editing, R.C. and A.D.F.; funding acquisition, R.C., F.A., M.L., G.D., S.M.S.P. and M.-C.C. All authors have read and agreed to the published version of the manuscript.

Funding: This research was funded by European Union's Horizon 2020 research and innovation program, under grant agreement no. 824957 ("BeforeHand:" Boosting Performance of Phase Change Devices by Hetero- and Nanostructure Material Design).

Conflicts of Interest: The authors declare no conflict of interest.

References

1. Bandyopadhyay, D.; Sen, J. Internet of Things: Applications and Challenges in Technology and Standardization. *Wirel. Pers Commun.* **2011**, *58*, 49–69. [CrossRef]
2. Cappelletti, P.; Annunziata, R.; Arnaud, F.; Disegni, F.; Maurelli, A.; Zuliani, P. Phase Change Memory for Automotive Grade Embedded NVM Applications. *J. Phys. D Appl. Phys.* **2020**, *53*, 193002. [CrossRef]
3. Wuttig, M.; Yamada, N. Phase-Change Materials for Rewriteable Data Storage. *Natur. Mater.* **2007**, *6*, 824–832. [CrossRef] [PubMed]
4. Fantini, A.; Perniola, L.; Armand, M.; Nodin, J.F.; Sousa, V.; Persico, A.; Cluzel, J.; Jahan, C.; Maitrejean, S.; Lhostis, S.; et al. Comparative Assessment of GST and GeTe Materials for Application to Embedded Phase-Change Memory Devices. In Proceedings of the 2009 IEEE International Memory Workshop, Monterey, CA, USA, 10–14 May 2009; pp. 1–2.
5. Hubert, Q.; Jahan, C.; Toffoli, A.; Navarro, G.; Chandrashekar, S.; Noe, P.; Blachier, D.; Sousa, V.; Perniola, L.; Nodin, J.-F.; et al. Lowering the Reset Current and Power Consumption of Phase-Change Memories with Carbon-Doped $Ge_2Sb_2Te_5$. In Proceedings of the 2012 4th IEEE International Memory Workshop, Milan, Italy, 20–23 May 2012; pp. 1–4.
6. Li, T.; Shen, J.; Wu, L.; Song, Z.; Lv, S.; Cai, D.; Zhang, S.; Guo, T.; Song, S.; Zhu, M. Atomic-Scale Observation of Carbon Distribution in High-Performance Carbon-Doped $Ge_2Sb_2Te_5$ and Its Influence on Crystallization Behavior. *J. Phys. Chem. C.* **2019**, *123*, 13377–13384. [CrossRef]
7. Kikuchi, S.; Oh, D.Y.; Kimura, I.; Nishioka, Y.; Ueda, M.; Endo, M.; Kokaze, Y.; Suu, K. Preparation of Oxygen-Doped and Nitrogen-Doped Ge-Sb-Te System Thin Film for Phase Change Random Access Memory by RF Magnetron Sputtering. In Proceedings of the 2006 7th Annual Non-Volatile Memory Technology Symposium, San Mateo, CA, USA, 5–8 November 2006; pp. 81–83.
8. Privitera, S.; Rimini, E.; Zonca, R. Amorphous-to-Crystal Transition of Nitrogen- and Oxygen-Doped $Ge_2Sb_2Te_5$ Films Studied by in Situ Resistance Measurements. *Appl. Phys. Lett.* **2004**, *85*, 3044–3046. [CrossRef]
9. Lai, Y.; Qiao, B.; Feng, J.; Ling, Y.; Lai, L.; Lin, Y.; Tang, T.; Cai, B.; Chen, B. Nitrogen-Doped $Ge_2Sb_2Te_5$ Films for Nonvolatile Memory. *J. Elec. Mater.* **2005**, *34*, 176–181. [CrossRef]
10. Horii, H.; Yi, J.H.; Park, J.H.; Ha, Y.H.; Baek, I.G.; Park, S.O.; Hwang, Y.N.; Lee, S.H.; Kim, Y.T.; Lee, K.H.; et al. A Novel Cell Technology Using N-Doped GeSbTe Films for Phase Change RAM. In Proceedings of the 2003 Symposium on VLSI Technology. Digest of Technical Papers (IEEE Cat. No.03CH37407), Kyoto, Japan, 10–12 June 2003; pp. 177–178.
11. Zuliani, P.; Varesi, E.; Palumbo, E.; Borghi, M.; Tortorelli, I.; Erbetta, D.; Libera, G.D.; Pessina, N.; Gandolfo, A.; Prelini, C.; et al. Overcoming Temperature Limitations in Phase Change Memories With Optimized $Ge_xSb_yTe_z$. *IEEE Trans. Electron. Devices* **2013**, *60*, 4020–4026. [CrossRef]
12. Privitera, S.M.S.; López García, I.; Bongiorno, C.; Sousa, V.; Cyrille, M.C.; Navarro, G.; Sabbione, C.; Carria, E.; Rimini, E. Crystallization Properties of Melt-Quenched Ge-Rich GeSbTe Thin Films for Phase Change Memory Applications. *J. Appl. Phys.* **2020**, *128*, 155105. [CrossRef]

13. Redaelli, A.; Petroni, E.; Annunziata, R. Material and Process Engineering Challenges in Ge-Rich GST for Embedded PCM. *Mater. Sci. Semicond. Process.* **2022**, *137*, 106184. [CrossRef]
14. Tae Kim, Y.; Kim, S.-I. Comparison of Thermal Stabilities between Ge-Sb-Te and In-Sb-Te Phase Change Materials. *Appl. Phys. Lett.* **2013**, *103*, 121906. [CrossRef]
15. Morikawa, T.; Kurotsuchi, K.; Kinoshita, M.; Matsuzaki, N.; Matsui, Y.; Fujisaki, Y.; Hanzawa, S.; Kotabe, M.; Moriya, H.; Iwasaki, T.; et al. Doped In-Ge-Te Phase Change Memory Featuring Stable Operation and Good Data Retention. In Proceedings of the 2007 IEEE International Electron Devices Meeting, Washington, DC, USA, 10 December 2007; pp. 307–310.
16. Fallica, R.; Stoycheva, T.; Wiemer, C.; Longo, M. Structural and Electrical Analysis of In Sb Te-based PCM cells. Physica status solidi (RRL). *Rapid Res. Lett.* **2013**, *7*, 1009–1013. Available online: https://onlinelibrary.wiley.com/doi/abs/10.1002/pssr.201308074 (accessed on 20 March 2022).
17. Selmo, S.; Cecchini, R.; Cecchi, S.; Wiemer, C.; Fanciulli, M.; Rotunno, E.; Lazzarini, L.; Rigato, M.; Pogany, D.; Lugstein, A.; et al. Low Power Phase Change Memory Switching of Ultra-Thin $In_3Sb_1Te_2$ Nanowires. *Appl. Phys. Lett.* **2016**, *109*, 213103. [CrossRef]
18. Cecchini, R.; Selmo, S.; Wiemer, C.; Rotunno, E.; Lazzarini, L.; De Luca, M.; Zardo, I.; Longo, M. Single-Step Au-Catalysed Synthesis and Microstructural Characterization of Core–Shell Ge/In–Te Nanowires by MOCVD. *Mater. Res. Lett.* **2018**, *6*, 29–35. [CrossRef]
19. Chèze, C.; Righi Riva, F.; Di Bella, G.; Placidi, E.; Prili, S.; Bertelli, M.; Diaz Fattorini, A.; Longo, M.; Calarco, R.; Bernasconi, M.; et al. Interface Formation during the Growth of Phase Change Material Heterostructures Based on Ge-Rich Ge-Sb-Te Alloys. *Nanomaterials* **2022**, *12*, 1007. [CrossRef]
20. Nolot, E.; Sabbione, C.; Pessoa, W.; Prazakova, L.; Navarro, G. Germanium, Antimony, Tellurium, Their Binary and Ternary Alloys and the Impact of Nitrogen: An X-Ray Photoelectron Study. *Appl. Surf. Sci.* **2021**, *536*, 147703. [CrossRef]
21. Klein, A.; Dieker, H.; Späth, B.; Fons, P.; Kolobov, A.; Steimer, C.; Wuttig, M. Changes in Electronic Structure and Chemical Bonding upon Crystallization of the Phase Change Material $GeSb_2Te_4$. *Phys. Rev. Lett.* **2008**, *100*, 1–4. [CrossRef]
22. Baeck, J.H.; Ann, Y.; Jeong, K.-H.; Cho, M.-H.; Ko, D.-H.; Oh, J.-H.; Jeong, H. Electronic Structure of Te/Sb/Ge and Sb/Te/Ge Multi Layer Films Using Photoelectron Spectroscopy. *J. Am. Chem. Soc.* **2009**, *131*, 13634–13638. [CrossRef]
23. Kim, J.-J.; Kobayashi, K.; Ikenaga, E.; Kobata, M.; Ueda, S.; Matsunaga, T.; Kifune, K.; Kojima, R.; Yamada, N. Electronic Structure of Amorphous and Crystalline $(GeTe)_{1-x}(Sb_2Te_3)_x$ Investigated Using Hard X-ray Photoemission Spectroscopy. *Phys. Rev. B* **2007**, *76*, 115124. [CrossRef]
24. Lee, Y.M.; Jung, M.-C.; Shin, H.J.; Kim, K.; Song, S.A.; Jeong, H.S.; Ko, C.; Han, M. Temperature-Dependent High-Resolution X-Ray Photoelectron Spectroscopic Study on $Ge_1Sb_2Te_4$. *Thin Solid Films* **2010**, *518*, 5670–5672. [CrossRef]
25. Sosso, G.C.; Caravati, S.; Gatti, C.; Assoni, S.; Bernasconi, M. Vibrational Properties of Hexagonal $Ge_2Sb_2Te_5$ from First Principles. *J. Phys. Condens. Matter* **2009**, *21*, 245401. [CrossRef]
26. Abou El Kheir, O.; Dragoni, D.; Bernasconi, M. Density Functional Simulations of Decomposition Pathways of Ge-Rich GeSbTe Alloys for Phase Change Memories. *Phys. Rev. Mater.* **2021**, *5*, 095004. [CrossRef]
27. Abou El Kheir, O.; Bernasconi, M. High-Throughput Calculations on the Decomposition Reactions of Off-Stoichiometry GeSbTe Alloys for Embedded Memories. *Nanomaterials* **2021**, *11*, 2382. [CrossRef]
28. Zuliani, P.; Palumbo, E.; Borghi, M.; Dalla Libera, G.; Annunziata, R. Engineering of Chalcogenide Materials for Embedded Applications of Phase Change Memory. *Solid-State Electron.* **2015**, *111*, 27–31. [CrossRef]
29. Bragaglia, V.; Holldack, K.; Boschker, J.E.; Arciprete, F.; Zallo, E.; Flissikowski, T.; Calarco, R. Far-Infrared and Raman Spectroscopy Investigation of Phonon Modes in Amorphous and Crystalline Epitaxial $GeTe-Sb_2Te_3$ Alloys. *Sci. Rep.* **2016**, *6*, 28560. [CrossRef]
30. Kumar, A.; Cecchini, R.; Wiemer, C.; Mussi, V.; De Simone, S.; Calarco, R.; Scuderi, M.; Nicotra, G.; Longo, M. Phase Change Ge-Rich Ge–Sb–Te/Sb_2Te_3 Core-Shell Nanowires by Metal Organic Chemical Vapor Deposition. *Nanomaterials* **2021**, *11*, 3358. [CrossRef]
31. Di Biagio, F.; Cecchi, S.; Arciprete, F.; Calarco, R. Crystallization Study of Ge-Rich $(GeTe)_m(Sb_2Te_3)_n$ Using Two-Step Annealing Process. *Phys. Status Solidi RRL* **2019**, *13*, 1800632. [CrossRef]
32. Lannin, J.S.; Maley, N.; Kshirsagar, S.T. Raman scattering and short range order in amorphous germanium. *Solid State Commun.* **1985**, *53*, 939–942. [CrossRef]
33. Kazimierski, P.; Tyczkowski, J.; Kozanecki, M.; Hatanaka, Y.; Aoki, T. Transition from Amorphous Semiconductor to Amorphous Insulator in Hydrogenated Carbon−Germanium Films Investigated by Raman Spectroscopy. *Chem. Mater.* **2002**, *14*, 4694–4701. [CrossRef]
34. Jamali, H.; Mozafarinia, R.; Eshaghi, A. The Effect of Carbon Content on the Phase Structure of Amorphous/Nanocrystalline $Ge_{1-x}C_x$ Films Prepared by PECVD. *Surf. Coat. Technol.* **2017**, *310*, 1–7. [CrossRef]
35. Prazakova, L.; Nolot, E.; Martinez, E.; Rouchon, D.; Fillot, F.; Bernier, N.; Elizalde, R.; Bernard, M.; Navarro, G. The Effect of Ge Content on Structural Evolution of Ge-Rich GeSbTe Alloys at Increasing Temperature. *Materialia* **2022**, *21*, 101345. [CrossRef]
36. Cecchi, S.; Lopez Garcia, I.; Mio, A.M.; Zallo, E.; Abou El Kheir, O.; Calarco, R.; Bernasconi, M.; Nicotra, G.; Privitera, S.M.S. Crystallization and Electrical Properties of Ge-Rich GeSbTe Alloys. *Nanomaterials* **2022**, *12*, 631. [CrossRef] [PubMed]
37. Goriparti, S.; Miele, E.; Scarpellini, A.; Marras, S.; Prato, M.; Ansaldo, A.; DeAngelis, F.; Manna, L.; Zaccaria, R.P.; Capiglia, C. Germanium Nanocrystals-MWCNTs Composites as Anode Materials for Lithium Ion Batteries. *ECS Trans.* **2014**, *62*, 19–24. [CrossRef]

38. Sousa, V.; Navarro, G.; Castellani, N.; Coue, M.; Cueto, O.; Sabbione, C.; Noe, P.; Perniola, L.; Blonkowski, S.; Zuliani, P.; et al. Operation Fundamentals in 12Mb Phase Change Memory Based on Innovative Ge-Rich GST Materials Featuring High Reliability Performance. In Proceedings of the 2015 Symposium on VLSI Technology (VLSI Technology), Kyoto, Japan, 16–18 June 2015; pp. T98–T99.
39. Kiouseloglou, A.; Navarro, G.; Sousa, V.; Persico, A.; Roule, A.; Cabrini, A.; Torelli, G.; Maitrejean, S.; Reimbold, G.; De Salvo, B.; et al. A Novel Programming Technique to Boost Low-Resistance State Performance in Ge-Rich GST Phase Change Memory. *IEEE Trans. Electron. Devices* **2014**, *61*, 1246–1254. [CrossRef]

Article

Phase Separation in Ge-Rich GeSbTe at Different Length Scales: Melt-Quenched Bulk versus Annealed Thin Films

Daniel Tadesse Yimam [1], A. J. T. Van Der Ree [1], Omar Abou El Kheir [2], Jamo Momand [1], Majid Ahmadi [1], George Palasantzas [1], Marco Bernasconi [2] and Bart J. Kooi [1,*]

[1] Zernike Institute for Advanced Materials, University of Groningen, Nijenborgh 4, 9747 AG Groningen, The Netherlands; d.t.yimam@rug.nl (D.T.Y.); a.j.t.van.der.ree@rug.nl (A.J.T.V.D.R.); j.momand@gmail.com (J.M.); majid.ahmadi@rug.nl (M.A.); g.palasantzas@rug.nl (G.P.)

[2] Department of Materials Science, University of Milano-Bicocca, Via R. Cozzi 55, I-20125 Milano, Italy; o.abouelkheir@campus.unimib.it (O.A.E.K.); marco.bernasconi@unimib.it (M.B.)

* Correspondence: b.j.kooi@rug.nl

Abstract: Integration of the prototypical GeSbTe (GST) ternary alloys, especially on the GeTe-Sb_2Te_3 tie-line, into non-volatile memory and nanophotonic devices is a relatively mature field of study. Nevertheless, the search for the next best active material with outstanding properties is still ongoing. This search is relatively crucial for embedded memory applications where the crystallization temperature of the active material has to be higher to surpass the soldering threshold. Increasing the Ge content in the GST alloys seems promising due to the associated higher crystallization temperatures. However, homogeneous Ge-rich GST in the as-deposited condition is thermodynamically unstable, and phase separation upon annealing is unavoidable. This phase separation reduces endurance and is detrimental in fully integrating the alloys into active memory devices. This work investigated the phase separation of Ge-rich GST alloys, specifically $Ge_5Sb_2Te_3$ or GST523, into multiple (meta)stable phases at different length scales in melt-quenched bulk and annealed thin film. Electron microscopy-based techniques were used in our work for chemical mapping and elemental composition analysis to show the formation of multiple phases. Our results show the formation of alloys such as GST213 and GST324 in all length scales. Furthermore, the alloy compositions and the observed phase separation pathways agree to a large extent with theoretical results from density functional theory calculations.

Keywords: phase change materials; Ge-rich GST; pulsed laser deposition; phase separation; GGST; EDX elemental chemical mapping; embedded memory; density functional theory

1. Introduction

Phase change materials have been heavily investigated for many applications in fields of phase-change memories (PCMs), nanophononics, and neuromorphic applications for the past few years [1,2]. Although there have been many promising binary and ternary alloys with phase switching properties, by far the most studied and integrated phase change materials are the Ge-Sb-Te alloys, specifically $Ge_2Sb_2Te_5$ or GST225. Despite being the center of research in the field with attractive crystallization properties and stability and being the front runner up for future memory and optoelectronic devices, GST225 is by no means without its limitations. One main reliability issue, especially crucial for PCMs, is the gradual "drift" of resistance in the metastable amorphous phase of GST225 over time [3–5]. The resistivity drift towards higher values, due to structural relaxations to a thermodynamically stable state [6], is particularly detrimental when going beyond two memory states per bit, e.g., for neuromorphic computing. Other issues, such as void formation and electromigration, have also been reported [7,8]. In addition, the low crystallization temperature of GST225 (≈150 °C) limits the material's potential for automotive and aeronautics applications [9]. Simply put, GST225 could not meet the automotive specifications and soldering threshold needed in embedded memory applications [10,11].

There has been a growing trend in recent years in shifting away from pure GST225 alloy for PCMs to find a better alternative with high crystallization temperature and thermal stability. One way of achieving the needed properties is by "doping" pure GST225 with impurities [12,13]. Incorporating impurities into GST225 induced higher crystallization temperature, better structural and thermal stability, and faster switching [8,14]. Interestingly, among all doping impurities, nitrogen doping is by far the most attractive and results in promising experimental works in the field [15–20]. Another approach is to deviate in composition from the stable GST225 phase by incorporating excess Ge, creating Ge-rich GST alloys. Increasing the Ge content in the GST alloys shows an increase in the crystallization temperature, promoting high data retention and endurance [10,21,22]. Moreover, in addition to Ge enriching, nitrogen doping of Ge-rich GST alloys has also been studied for increased thermal stability and higher crystallization temperatures, which are attractive for future memory devices [9,23,24].

A common similarity in almost all Ge-rich GST alloys is thermodynamically unfavorable initial phases. Although research on the alloys produced many promising results, phase separation into stable phases and local compositional variation upon melt-quenching could not be avoided [9,25–27]. This issue poses a significant problem for programming operations in an active device. The local composition variation and phase separation into Ge-rich and Sb-rich regions compromise the device's functionality [10,22,28,29]. Moreover, the excess Ge content in Ge-rich alloys increases oxidation susceptibility with lowered crystallization temperature and Te enrichment [9,30]. This work reports experimental results of local phase separation in a Ge-GST alloy GST523. The choice of the specific starting composition, i.e., GST523, is due to previous experimental and theoretical results on Ge-rich GST alloys on the Ge-Sb_2Te_3 tie-line [31,32]. Among a series of Ge-rich GST, GST523 shows a high crystallization temperature, while the Ge content is still moderately low. Therefore, the alloy's phase separation into GSTxyz and pure Ge is still limited. In addition, GST523 is an interesting alloy to investigate, since it is on the Ge-Sb_2Te_3 tie-line and can also be potentially produced by alternating thin Ge and Sb_2Te_3 layers. We provide evidence for the formation of multiple phases upon melting and quenching of a thermodynamically unstable Ge-rich GST phase at different length scales. A large area morphology and elemental composition of a Ge-rich GST sample were analyzed using scanning electron microscopy (SEM). In addition, we use scanning/transmission electron microscopy (S/TEM) for composition analysis of pulsed laser deposited Ge-rich GST alloys on a smaller length scale. The experimental results have been compared with calculations based on Density Functional Theory (DFT) on the thermodynamics of the decomposition pathways.

2. Experimental Methods

We prepared thin films of Ge-rich GST, GST225, and Sb_2Te_3 using pulsed laser deposition (PLD), with a KrF excimer laser operating at 248 nm wavelength. For GST225 and Sb_2Te_3 depositions, powder-sintered targets from K-TECH were used. For the deposition of Ge-rich GST thin films, an in-house target was made. Exact constituents of high-purity Ge, Sb, and Te atomic portions were sealed in a vacuum quartz ampule. The ampule was put into an oven and melted by gradually increasing the temperature to 950 °C and kept for 2 h. Finally, the ampule was cooled by water and crushed into a powder material. The ingot was ball milled for 5 h to produce a fine powder, which was then used to produce pellets of 20 mm diameter by a pressure press. Finally, the pellets were sintered at 300 °C for 1.5 h to produce a dense powder target with a composition of GST523. A fluence of 0.8 J cm^{-2}, processing gas (Ar) pressure of 0.12 mbar, and target-substrate distance of 55 mm were used for all depositions. The depositions were performed at room temperature to produce as-deposited amorphous samples. Reflection high-energy electron diffraction (RHEED) was used as an initial characterization for the amorphous nature of deposited thin films. Thin films were deposited on Si wafer covered with thermal SiO_2 film for ellipsometry analysis and on continuous carbon and Si_3N_4 TEM grids for scanning/transmission electron microscopy (S/TEM) imaging.

For scanning electron microscopy (SEM), a FEI Helios G4 CX and a FEI NovaNanoSEM 650 (both from Thermo Fisher – FEI, Hillsboro, OR, USA) equipped with an energy dispersive X-ray (EDX) detector were used to analyze the surface morphology of the initial crystalline ampule and the final powder-sintered target used for deposition. A Themis Z S/TEM (Thermo Fisher – FEI, Hillsboro, OR, USA) operating at 300 kV and a JEOL 2010 TEM (JEOL USA Inc., Boston, MA, USA) operating at 200 kV were used for plan-view imaging and elemental analysis of deposited thin films. A cross-sectional specimen of the as-deposited and crystalline Ge-rich GST thin films was prepared with a focused ion beam (FIB) (FEI Helios G4 CX).

Dynamic ellipsometry measurements (DE) were conducted to investigate the phase transformations of the as-deposited amorphous thin films. A heating stage (HTC-100), attached to a J. Woollam UV-VIS variable angle spectroscopic ellipsometer (VASE), was used for the measurements. The ramp and hold steps of the measurements were controlled by the TempRampVASE software (Version 1.06), while the WVASE software (Version 3.916) monitored the measurement intensities. All DE measurements were conducted in air at a 70° incidence angle and with a 5 °C min^{-1} heating rate. Spectroscopic ellipsometry (SE) measurements were performed on the thin films before and after heating. Measurements were conducted in the spectral range of 300–1700 nm. For fitting ease, maximum accuracy, and reduction of parameter correlation, measurements at three angles of incidence (65°, 70°, and 75°) were collected. For refractive index (n) and extinction coefficient (k) extraction, measurement data were fitted with the Tauc–Lorentz oscillator model using the WVASE fitting software.

3. Results and Discussion

To investigate the phase separation of a relatively large liquid volume of Ge-rich GST alloy, we produced a GST523 ingot heated to 950 °C for a sufficiently long time to ensure that a completely homogeneous GT523 liquid is obtained. The quartz tube is rapidly quenched into water by pulling it out of the furnace directly into a water bath at room temperature. The strong driving force for phase separation of GST523 combined with the relatively fast cooling produces phase separation on the scale of 1–100 μm which can be very well assessed by SEM and also EDX performed in the SEM. Figure 1a shows a backscatter image with Z contrast and, therefore, the observed dark phase is Ge-rich and the observed brighter phase in the image is Sb-rich. The EDX results in Figure 1b–e show that primary Ge dendrites are formed. The remaining liquid solidifies with relatively homogeneous Te concentration but with complementary Ge and Sb concentrations, i.e., phase separation in regions (1) high in Ge and low in Sb and (2) low in Ge and high in Sb. In extreme form, these two phases could be GeTe and Sb_2Te_3, respectively, but EDX shows that this is incorrect. Thus, the observed morphology and EDX results are also consistent; i.e., (pure) Ge forms the largest primary dendrites, because it is the phase with the highest melting temperature that solidifies first; then the Ge-rich GST phase solidifies as smaller secondary dendrites, because it has an intermediate melting temperature; and, finally, the remaining Sb-rich GST solidifies last, since it has the lowest melting temperature. The morphology of the phases observed in SEM (also higher magnification images than shown here), the large scale on which phase separation has occurred, and the well-known fact that GST alloys are very poor glass formers make it inevitable that all three distinct regions are crystalline in nature.

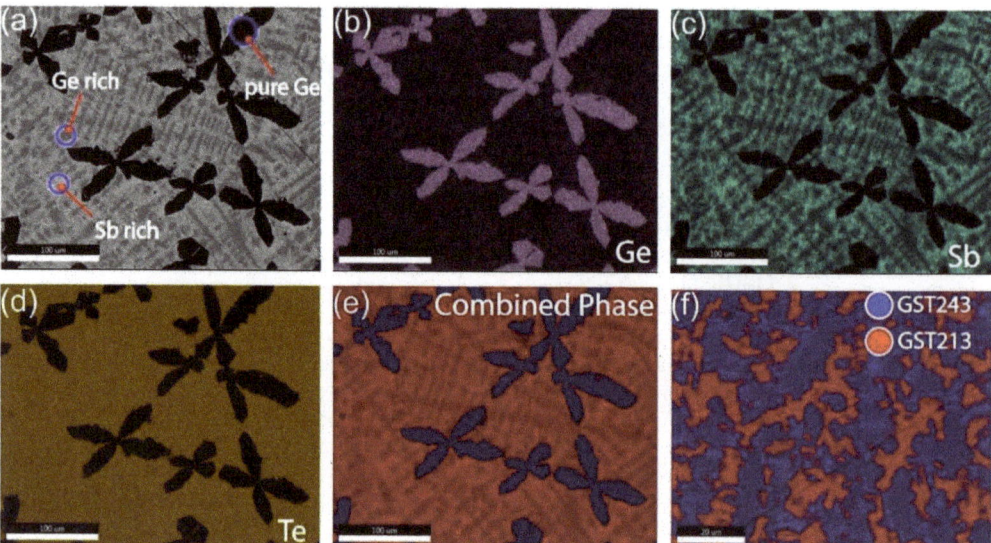

Figure 1. SEM–EDX mapping of water-quenched crystalline alloy showing the formation of multiple phases. (**a**) SEM image showing the formation of Ge dendrites from the excess germanium in the system and other phases. (**b**) Ge, (**c**) Sb, and (**d**) Te elemental maps. (**e**) A combined (Ge, Sb, and Te) elemental mapping with clear color contrast shows three phases. (**f**) When taken away from the pure Ge dendrites, an elemental map shows two distinct phases. The scale bar in (**a**–**e**) is 100 µm and in (**f**) is 20 µm.

Detailed quantitative analysis of SEM-EDX maps shows that the following three "phases" are formed: Ge, GST213, and GST243. Here, the term "phases" is associated with the local compositional variations observed from the experimental results. It also holds for the STEM-EDX results presented below. However, keep in mind that the local compositions can describe a real stable crystalline phase, such as pure Ge, but also metastable phases observed outside the GeTe-Sb_2Te_3 tie-line. In case of water-quenching a relatively bulky liquid sample present in evacuated quartz tube, the cooling rates are relatively low, and the crystalline phases formed are similar to the trigonal stable GST phases observed in ref. [33]. Nevertheless, in this respect, GST213 can be considered with GeTe as starting phase, where about one third of the Ge is replaced by Sb, and GST243 as a Ge-rich and Te-poor GST with excess Sb. The overall composition for the GST523 sample measured for the EDX map shown in Figure 1e gave on average (in at.%) 47.1 Ge, 21.8 Sb, and 31.1 Te (with quantification error in at.% of 2.7 for Ge, 1.4 for Sb, and 1.4 for Te), which is reasonably close to the intended one. Additional information on the SEM–EDX analysis is presented in Supplementary Materials Section S1. The observed deviation from GST523 probably originates from the coarseness of the primary Ge dendrites. Then, even for a large area, such as that shown in Figure 1a, considerable fluctuations can occur depending on how much of the Ge dendrites are exactly in the analyzed area. Since we know the overall composition, which is GST523, and we know reasonably the composition of the three phase-separated phases (Ge, GST213, and GST243), we can estimate the fraction of atoms present in each of the three phases. Then 30% of the atoms are in Ge, 40% in GST213, and 30% in GST243. When, for simplicity, assuming that the atomic densities are identical in the three phases (which of course is a relatively crude approximation), then the same three fractions can hold for the volume fraction of the three phases in the material and, when sufficiently randomly distributed, also the area fraction in Figure 1a. When looking at Figure 1a qualitatively, these fractions appear quite reasonable. However, the

EDX mapping provides quantitative output regarding the area fraction of the different phases. For example, Figure 1f quantifies the fractions of the GST213 (secondary dendritic phase) and GST243 as 60% and 40%, respectively, which matches well the 4:3 ratio estimated above.

In PLD, crystalline and amorphous (dense) targets can be used to ablate materials with a high-power laser and create thin-film layers on a given substrate. In another way, and relatively common in PLD, powdered sintered (porous) targets can be used as target materials for ablation. Combining individual elements of the necessary components and milling them into a fine powder, which is then pressed into a pellet and sintered in a furnace, will produce a usable target material. Of course, when dealing with the accurate weight of individual elements, there is always a risk of deviating from the correct final composition due to the loss of elements in the ball milling and mixing processes. Therefore, in the present work, multiple routes have been investigated to produce Ge-rich GST thin films. Since Ge-rich GST alloys can be considered Ge addition to Sb_2Te_3, GST523 can be seen as $(Ge)5 + (Sb_2Te_3)1$. Based on this, we produced "superlattice-like" heterostructures with alternating pure Ge and Sb_2Te_3 layers. The details of Ge target production and heterostructure depositions are explained in Supplementary Materials Section S1. Although the desired composition was achieved by varying the individual layer thicknesses, the produced films showed severe problems with Ge oxidation and delamination (see Supplementary Materials Section S2 (Figures S1 and S2). Another alternative was to initiate the deposition process directly from the crystalline target. Along the $GeTe-Sb_2Te_3$ tie-line, $Ge_xSb_yTe_z$ ternary alloys have multiple known stable phases, with minimum or no phase separation (see Figure 2a inset). However, when Ge content is increased and deviates away from this tie-line, the stability of the alloy is thermodynamically unfavorable, and phase separation is inevitable.

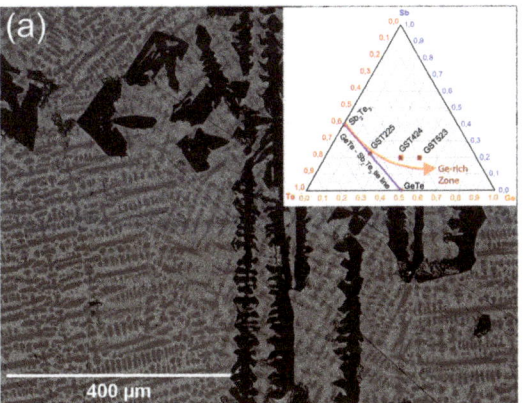 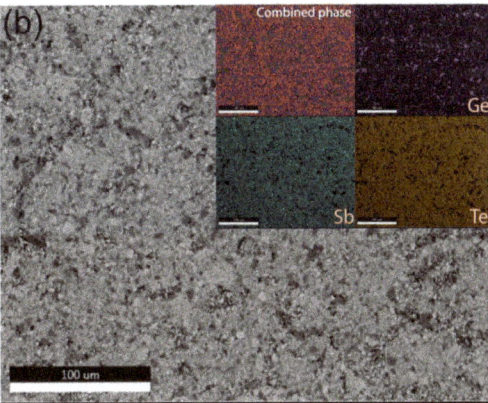

Figure 2. (a) GST523 crystalline target made by melting a combination of constituent elements at 950 °C. Clear phase separation is visible from the image. Inset shows a phase diagram of Ge-Sb-Te alloys indicating the tie-line of Sb_2Te_3-GeTe and the deviation toward Ge-rich GST regions. (b) Powder-sintered target, prepared by crushing the crystalline target for better yield and stoichiometric transfer. Inset shows SEM EDX mapping of the powder-sintered target. A "single phase"-like powder target of GST523 was produced.

Figure 2a shows a backscatter SEM image of a (dense) crystalline GST523 PLD target with clearly distinguishable phases present. The laser spot size for most PLD set-ups is much larger (3–9 mm^2) than the segregated individual grains of 10–100 µm range. Therefore, we could assume some form of homogeneity inside the laser–target interaction volume during ablation. However, in our deposition using the crystalline GST523 target depicted in Figure 2a, the thin-film composition deviated significantly from the initial target stoichiometry. One primary reason is that the ablation threshold and characteristics are different for individual phases, with different elemental compositions present in the target material. As a next alternative, we crushed the crystalline ingot into a fine powder, then compacted it into a pellet using high pressure. The pellet was sintered at an elevated temperature in the final step. Figure 2b shows an SEM image of the final powder-sintered target, and the large area SEM-EDX analysis shows the homogeneous distribution of the elemental constituents in the inset. Although still with an overall GST523 composition, we created in this way a "pseudo-single phase" powder target for deposition. Thin films of Ge-rich GST were deposited, and the stoichiometry transfer and the film quality were dramatically improved when going from the initial crystalline target (Figure 2a) to the powder-sintered target (Figure 2b).

Once a suitable target was made for material ablation, multiple depositions were performed to create amorphous as-deposited Ge-rich GST samples. Depending on the technique used, the as-deposited phase of PCMs have different structures and properties. It is especially crucial when considering the thermal stability of the produced material after melt-quenching. Therefore, it is ideal to start with an as-deposited phase that closely resembles the melt-quenched phase structure. Although attractive for large-scale production with usually a homogeneous chemical composition in the as-deposited phases, sputtering techniques often produce samples with reduced thermal stability after melt-quenching [34]. On the other hand, pulsed laser deposited samples should have structures already more closely resembling phases after melt-quenching. One piece of evidence for this is the presence of nanocrystals embedded in the amorphous matrix of pulsed laser deposited thin films (especially Sb_2Te_3). However, it is not desirable to have a crystalline phase inside the amorphous matrix initially at room temperature deposition, but this also means that the as-deposited phase will not change dramatically after the melt-quench process, which is also known to produce crystalline embryos embedded in an amorphous matrix [35,36].

The variation of optical properties between amorphous and crystalline phases in phase change materials can be probed continuously to monitor the phase transition upon heating. Reflectance data are continuously monitored in dynamic ellipsometry (DE) measurements while the sample is heated with a constant temperature ramping. We have also prepared thin films of Sb_2Te_3 and GST225 for DE measurements for better insight into crystallization temperature variations with Ge content. Spectroscopic ellipsometry measurements and data fittings are also given in Supplementary Materials Section S4 (Figure S4). Figure 3a shows the results from DE measurement for Sb_2Te_3, GST225, and Ge-rich GST thin films. All films have thicknesses of 35–40 nm, and a heating rate of 5 °C min^{-1} was used. The normalized intensity shows the measured value for the ellipsometry parameter ψ at 1400 nm wavelength. The figure shows abrupt changes in the measured parameter upon phase transformation for all three phases. In Figure 3b, the first derivatives of the normalized measurement values are given. By fitting the curves with a Gaussian function, we can accurately extract the crystallization temperature for the thin films. Sb_2Te_3 and GST225 films crystallize at 120 °C and 146 °C, respectively. However, for Ge-rich GST thin films, crystallization happens at a higher temperature of 213 °C. Our Ge-rich GST thin films, thus, show Tx values that are about 70 °C higher than that of GST225 thin films. Given the excess Ge content present in the Ge-rich GST thin films, the relatively higher Tx is not surprising. It has been shown that increasing Ge content leads to higher Tx values for the ternary phase [10,31].

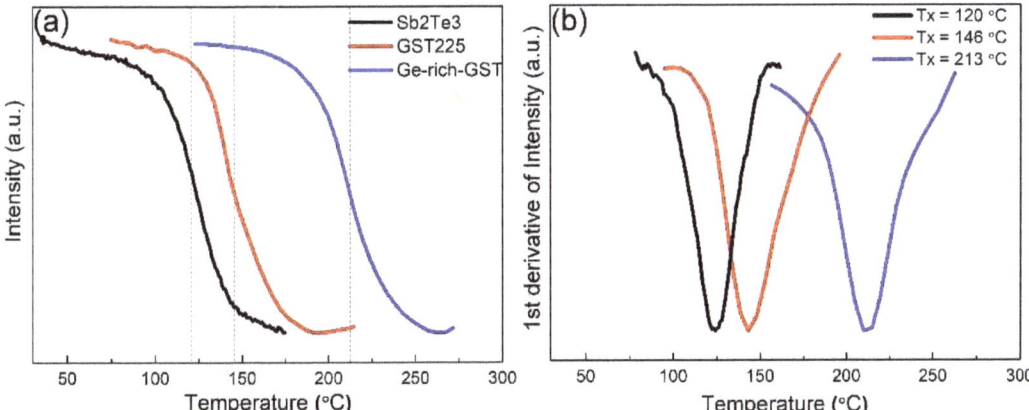

Figure 3. (a) Crystallization curves from dynamic ellipsometry measurements. Measurements of Sb$_2$Te$_3$ and GST225 are also plotted, in addition to the Ge-rich GST thin film, for better comparison. An abrupt change in intensity indicates a phase transformation upon heating. (b) The first derivative of the intensity from the dynamic ellipsometry measurement indicates exact crystallization temperature. A higher Tx value is observed for Ge-rich GST samples.

Crystallization dynamics studies on PCM thin films, with and without capping layers, showed the effect of oxidation on the crystallization onset. In general, uncapped films show transition temperatures lower than capped thin films. The reduction in the phase transition temperature is associated with heterogeneous nucleation at the oxidized regions of the uncapped thin films, which has been shown in crystallization dynamics works on prototypical PCMs such as GST225 and GeTe [37–39]. The effect worsens for the Ge-rich GST thin films containing excess elemental Ge with high oxidation affinity. The Ge migration towards the top surface (as evident in Figure 5a,c) contributes heavily to the phase separation, creating Ge-depleted regions in the lower parts of the thin film. The Ge depleted regions might have a transition temperature lower than other phases present in the thin film. It is worth mentioning here that we have to be cautious when using the term "crystallization temperature" for Ge-rich GST alloys. Given the complexity of the crystallization dynamics and the phase separations in the samples, it might be wise to use the term "transition temperature".

TEM-EDX analysis of the as-deposited Ge-rich GST thin films shows an average composition (in at.% and with atomic error <2% for all elements) of 46 Ge, 22 Sb, and 32 Te, which deviates from the initial target composition GST523. This deviation is caused by the formation of Ge-rich particulates (see Supplementary Materials Section S3 Figure S3). Formation of particulates is inherent to most PLD systems. Figure 4 shows cross-sectional images and EDX results of as-deposited and annealed Ge-rich GST samples obtained using TEM and S/TEM characterization. Figure 4a,b show BF-TEM images of a Ge-rich GST thin film before crystallization and after annealing at 450 °C for 30 min. Brighter and darker areas are visible in the as-deposited phase, and the image contrast is not homogeneous. The contrast is attributed to the local variation in composition and not because of diffraction contrast [26]. After annealing, the local composition is disturbed due to clusters of grain formations with different compositions. The first reasonable assumption for the image contrast would be that, since Sb and Te have comparable atomic numbers, Ge-rich and Ge-poor regions are present. This assumption is then verified by the high-angle annular dark-field (HAADF)-STEM images and STEM-EDX chemical mappings. Figure 4c shows the HAADF image for the annealed Ge-rich GST sample with clear contrast with local composition variations. Figure 4d–f shows the chemical mapping distribution of Ge, Sb, and Te, respectively. Comparing the HAADF image with the EDX mappings, it is clear that the darker regions in the image indeed represent Ge-rich zones. Although some local variation in all constituents is present, the Sb composition shows a relatively homogeneous map compared to Ge and Te. The homogeneous Sb intensity is contradictory to the results found in the SEM-EDX analysis of the melt-quench crystalline target, where Te was relatively homogeneous. This discrepancy could be attributed to the length scale associated with the measurement areas. The measurement and analysis are in the nm range for the thin film and the μm range for the crystalline target. Underlying is that the length scales in the crystalline targets are associated with quenching relatively slowly from the melt at high temperature, whereas the length scales in the thin film are associated with heating the solid (initially amorphous) phase. Of course, decomposition can occur much faster over long distances in liquid at high temperature than in solid at low temperature.

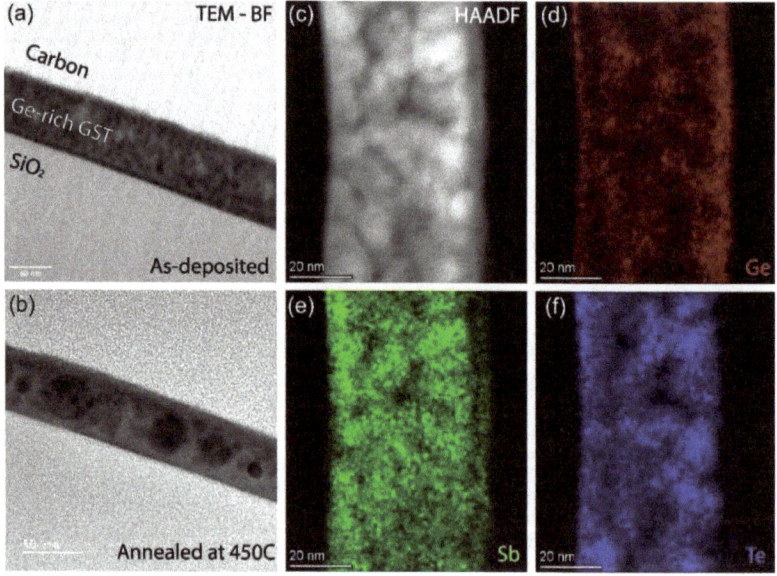

Figure 4. BF-TEM images of Ge-rich GST thin film in (**a**) as-deposited phase and (**b**) after annealing at 450 °C for 30 min. High-angle annular dark-field (HAADF)-STEM image (**c**) and STEM-EDX chemical compositions of (**d**) Ge, (**e**) Sb, and (**f**) Te for the annealed Ge-rich GST sample.

To further investigate the phase decomposition and formation of multiple phases in the Ge-rich GST thin films and compare the STEM-EDX results with the SEM-EDX analysis of the crystalline target, we need to analyze the local composition variation in the thin film. A somewhat crude way of investigating the local chemical variations is by analyzing the color contrast in the combined chemical mapping, in addition to the intensity line profiles. The combined chemical mapping and the selected areas for local composition analysis are given in Figure 5a. The line intensities of all elements (presented in Figure 5c) change throughout the film thickness, indicating the presence of multiple regions with different compositions and phase separations. The phase separations are associated with the Ge-rich phases and Te/Sb-rich regions. As can be seen from the line profiles, an increase in Ge intensity is accompanied by a decrease in Te/Sb intensity and vice versa. What is interesting is the migration of Ge to the film surface. This Ge migration is strongly driven by oxidation, which pulls Ge atoms to the surface, and this is well observable in cross-sections of thin films such as in Figure 5, but we also observed this for GST225 nanoparticles, where after prolonged exposure to air at room temperature the particles with an initial homogeneous composition develop a clear Ge-oxide outer shell [40]. However, also below the top surface the Ge-rich areas appear oxidized, as there is a clear direct correlation between Ge and oxygen concentration (see Figure 5b,c). However, it is important to note that this oxidation occurs when the very thin TEM lamella produced by FIB is some time in air before it is inserted in the TEM. Therefore, this subsurface oxidation did not affect the phase separation process during the thermal anneal. Figure 5e shows a high-resolution STEM image of some of the grains found in the sample. Most of the grains in the image appear to be relatively pure Ge crystals (with d-spacing of 3.25 Å). Another grain (with a lower d-spacing of 1.84 Å) is also presented which could not be indexed to any known Ge or GST phase.

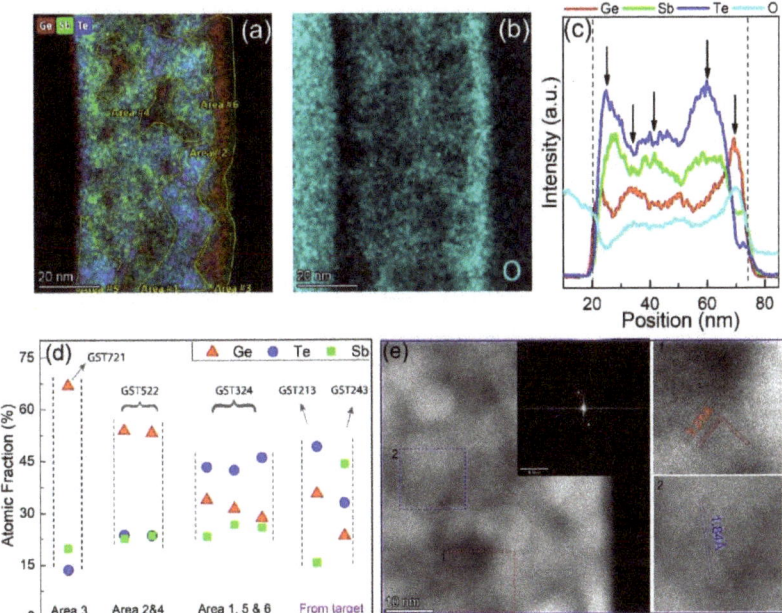

Figure 5. (a) The combined (Ge, Sb, and Te) STEM-EDX chemical mapping of the Ge-rich GST sample. Based on color contrast, multiple areas are selected for local composition analysis. The EDX chemical composition of oxygen is given in (b). (c) The line profile of the combined and the oxygen STEM-EDX chemical mapping of the Ge-rich GST sample. The oxygen line profile is scaled for better comparison with the Ge line profile. (d) Atomic fraction (in at.%) of the individual selected areas in (a) are plotted and compared with the atomic fraction from the melted target. (e) High-resolution images of some grains in the annealed sample with the inset showing the FFT.

A closer look at Figure 5a shows the presence of three primary color variations that stand out. This color variation is highly dependent on Ge content. What that means is that we can interpret the colors as high Ge content or "deep red" (near the film surface and represented by area #3), medium Ge content or "light red" (area #2 and #4), and low Ge content or "no red" (area #1, #5, and #6). Multiple areas from the different color groups have been selected, and their local composition is plotted in Figure 5d. For comparison, we also plotted the actual compositions (in at.%) of GST213 and GST243 phases identified and measured in the melt-quenched crystalline target. Note that the pure Ge phase found in the crystalline target is not included in the plot. Results from similar color groups have been averaged, and a composition for the tertiary alloy is extracted. As expected, the three regions show three different local compositions of GST721, GST522, and GST324. We can find multiple similarities and differences between the STEM-EDX and SEM-EDX results based on the plot in Figure 5d. The first and obvious similarity is the formation of pure Ge (close to pure for GST721) phase in the Ge-rich GST thin film and the melted crystalline target. The pure Ge presence is not surprising, since the initial composition has excess Ge. The SEM image in Figure 1a and the high-resolution STEM image in Figure 5e both confirm the presence of pure Ge. Another similarity is the presence of a Te-rich phase in both samples. GST324 in the annealed thin film and GST213 have comparable properties in the plot such that they both are higher in Te, similar in Ge, and lower in Sb compositions. One could even approximate the GST324 alloy as GST213 + GeTe.

Despite the similarities, some glaring differences exist between the compositions extracted from the thin film and from the crystalline target. The first is a Sb-rich phase in the crystalline target (GST243) which was not found in the thin film. Another difference is the phase GST522 found in the thin film, closely resembling the initial GST523 phase. It is worth noting that the difference in some of the local compositions of the thin film from the crystalline target could be attributed to multiple factors. One reason could be the difference in mobility of the constituents in the thin film and the melted target due to the supplied heat to the system. Starting from the melt, at 950 °C, would have the upper hand in providing enough energy to the system compared with annealing at 450 °C for 30 min. In addition, atomic mobilities are much higher in the liquid than in the solid. We could even argue that the GST522 phase is the prime example of the slow diffusion of constituents in the solid thin film, since it is close to the starting GST523 composition. Another factor for the difference could be the length scale used for the measurement. For the thin film, due to the high accelerating voltage of the S/TEM (300 kV compared to the 30 kV for SEM), the resolution in the chemical mapping is much better. Thus, the S/TEM-EDX provides an opportunity to correctly probe the local composition of the sample within nm scales. In the SEM-EDX, however, individual pixel sizes were limited to µm length scale for our sample. Therefore, the order of magnitude difference in resolution could produce aggregated/averaged compositions of nm length scale in a single µm length pixel, thus providing a biased estimation.

Nevertheless, it is expected that both techniques capture the correct length scale, since it is inherently longer by quenching, rather slowly (despite using water quenching), a bulk material from liquid at high temperature compared to heating a solid thin film at relatively low temperature. In a true memory device, of course, melt-quenching is also employed, but then, using short electrical pulses (<100 ns) and very small material volumes that are switched, melt-quenching can be extremely fast, thereby limiting phase separation. Nevertheless, Ge-rich GST alloys are very susceptible to phase separation, and even in such memory devices it is unlikely that it can be suppressed completely.

The phase separation of Ge-rich GST alloys has been theoretically investigated by means of DFT calculations to identify possible separation pathways [27,32,41]. To identify the similarities and differences between our result and that of the theoretically predicted phase separations, we plot the results from STEM-EDX and SEM-EDX analyses in the ternary phase diagram as presented in Figure 6a. The plot identifies four phases: 1—the starting phase, 2—the pure Ge phase, 3—phases close to the GeTe-Sb_2Te_3 tie-line, and, finally, 4—a Sb-rich phase found in the large scale SEM-EDX analysis of the melt-quenched crystalline target.

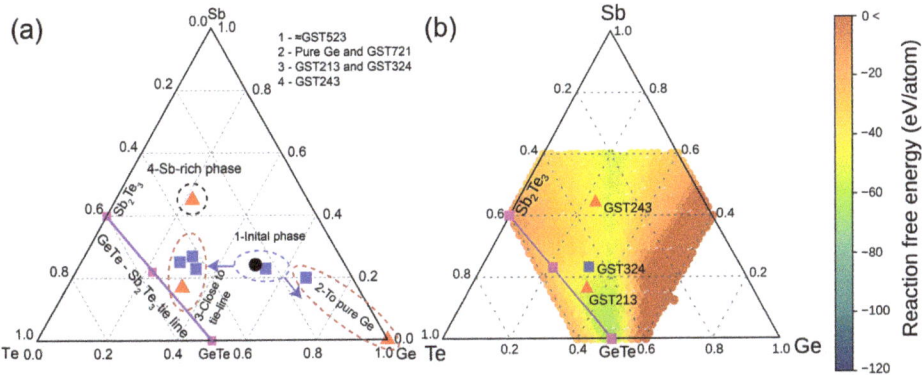

Figure 6. (a) The phase diagram of the starting composition and the final phases found from the STEM-EDX and SEM-EDX results. The black circle represents the initial phase of the thin film ((in at.%) 46 Ge, 22 Sb, and 32 Te). The blue squares represent elemental compositions extracted from STEM-EDX analysis of the annealed thin film. Finally, the red triangles show the compositions of phases found from the melt-quenched crystalline target in SEM-EDX analysis. The GeTe-Sb_2Te_3 tie-line, with Sb_2Te_3, GeTe, and GST225 compositions, are also presented for comparison. (**b**) Map of the decomposition pathways during crystallization of GST523 as based on DFT calculations (see ref. [27]). The color code gives the reaction free energy (meV/atom) to form the cubic alloys on the ternary phase diagram starting from the GST523 reactant. The same color code of the decomposition maps in ref. [27] for alloys on the Ge-GST124 tie-line is used.

When comparing our experimental results with previous DFT results for the GST523 alloy [32], both agree that Ge must be formed. Our experimental results do not show alloy formation on the GeTe-Sb_2Te_3 tie-line nor the formation of Sb_2Te_3, meaning that the phases seen are more likely metastable cubic and not trigonal. The DFT work shows that the formation of metastable cubic phases such as GST221 and GST323 became competitive when moving away from the trigonal phases. This observation seems to align with our experimental results, given the compositional similarities (GST323 from simulation to GST324 of STEM-EDX and GST213 of SEM-EDX results) and the energetically favorable decomposition into two phases (e.g., Ge and GST234) instead of three phases (e.g., Ge, GeTe, and GST221). Indeed, one apparent deviation from the theoretical results of ref. [32] is the presence of Sb-rich GST phase GST243 in our experimental result. We must, however, consider that only a subset of the possible decomposition pathways was analyzed in ref. [32]. In addition, Bordas et al. [42] provided an extensive analysis of the Ge-GeTe-Sb_2Te_3-Sb phase diagram, allowing the prediction of solidification paths and solidification sequences. For alloys in a very large region around the Ge-Sb_2Te_3 tie-line, the formation of pure Ge is inevitable during solidification. It is also predicted that forming a ternary phase close to pure Ge is highly unlikely, which we also report in our results. However, a major difference with the predictions of Bordas et al. occurs regarding the formed GSTxyz phases in addition to pure Ge. During solidification of alloys on the Ge-Sb_2Te_3 tie-line, the solidification steps according to Bordas et al. always contain alloys only found on the GeTe-Sb_2Te_3 tie-line. When cooling from high temperature they reported the formation of three alloys, GST225, GST124, and GST147, in different composition (and temperature) ranges. This deviates from our observation when cooling from the melt of GST523 alloy, where we observed the formation of GST213 and GST324 clearly deviating from the GeTe-Sb_2Te_3 tie-line. Nevertheless, the solidification sequence of, firstly, Ge and then, secondly, GST213 (resembling GeTe) and, thirdly and finally, GST243 (rather close to 'δ-Sb_2Te'), as deduced from the results related to Figure 1, is very much in line with the predictions of Bordas et al.

To improve over the previous theoretical analysis, we have here adopted the more systematic approach exploited in our previous work [27] to study the decomposition of alloys on the Ge-GST124 tie-line. The DFT formation free energy of the alloys in the metastable cubic phase was provided in ref. [27] for all compositions in the central part of the ternary phase diagram. The calculated free energy consists of the total energy at zero temperature and the configurational free energy (at room temperature) due to disorder in the cubic sublattices. We used this information to compute the reaction free energy for the decomposition pathways of GST523 into a generic $Ge_xSb_yTe_z$ cubic alloy plus elemental Ge, Sb, and Te and the binary GeTe and Sb_2Te_3 compounds, as discussed in ref. [27], to which we refer for all details. This calculation yields the map of the decomposition free energies shown in Figure 6b. Each point in the map gives the value of the decomposition free energy for the formation of the corresponding alloy from GST523. The vibrational contribution to the reaction free energy, not included in Figure 6b, was shown to be negligible [32]. The more negative the reaction free energy is, the more favored is the corresponding decomposition channel. The green regions in Figure 6b, thus, correspond to the more favorable decomposition products which do include the three alloys GST213, GST324, and GST243 seen experimentally.

The decomposition map in Figure 6b indicates that the formation of alloys close to GeTe should be thermodynamically favored as well. However, as discussed in ref. [32], the formation of alloys close to GeTe requires a strong segregation of both elemental Ge and Sb which might be kinetically more difficult. Kinetic hindrances might, thus, explain the absence of alloys close to the GeTe composition in the experimental samples.

4. Conclusions

In summary, chemical mappings and local phase separations at different length scales have been investigated in Ge-rich GST alloys using SEM-EDX and STEM-EDX techniques. For a bulk crystalline sample, water-quenching produced three thermodynamically favored phases. These phases are pure Ge, Ge-rich phase, and Sb-rich phases. In addition to the SEM-EDX measurements on a relatively large length scale, STEM-EDX analyses have been performed on a smaller length scale on annealed Ge-rich GST thin films. For deposition of initial Ge-rich GST as-deposited thin films, pulsed laser deposition parameters and target preparation steps were optimized. Dynamic ellipsometry measurements on the as-deposited Ge-rich GST thin films have been used to determine the crystallization temperatures accurately. Pulsed laser deposited Ge-rich GST thin films have an elemental composition close to GST523, and their crystallization temperature was 70 °C higher than holds for the prototypical GST225 alloys. STEM-EDX analysis on the annealed Ge-rich GST thin films shows that thermal treatment at 450 °C for 30 min induced phase separation into pure Ge and Ge-poor GST phases. One would assume that, because of thermodynamics, the phases observed after separation would lie on the GeTe-Sb_2Te_3 tie-line. Interestingly, that is not the case here. The identified phases actually lie in between the GeTe-Sb_2Te_3 and the Ge-Sb_2Te_3 tie-lines. This outcome was predicted in a previous theoretical work [32] to be due to the formation of the metastable cubic phases instead of the trigonal ones. Our DFT calculation of the reaction free energy provides a comprehensive map for the decomposition pathways of GST523 which encompass the compositions seen experimentally among those thermodynamically favored. Our work provides experimental and theoretical evidence to the possible separation pathways in Ge-rich GST alloys, which is a valuable input in choosing future Ge-rich GST alloys for phase change memory applications.

Supplementary Materials: The following supporting information can be downloaded at: https://www.mdpi.com/article/10.3390/nano12101717/s1, Figure S1: Transmission electron microscope image of a thin film produced by alternating sublayers of Ge and Sb_2Te_3. The number of pulses used to ablate the Ge and Sb_2Te_3 targets was relatively small to approximate the deposition as a "co-sputtering". The idea of creating an intermixed layer in the as-deposited amorphous phase was not successful, since the film was not stable, and the electron beam caused delamination. Figure S2: (a) High-angle annular dark-field (HAADF)-STEM image of a thin film cross-section produced by alternating Ge and Sb_2Te_3 layers. In (b–e) EDX mappings of the elements Ge, Sb, Te, and O are presented. (f) The line profiles of the different elements analyzed by STEM-EDX chemical mapping of the heterostructure. The oxygen line profile follows the Ge line profile, indicating the formation of GeO$_x$ layer. Figure S3: (a) A large area view of the Ge-rich GST thin film surface. Multiple particulates can be seen in all windows of the TEM grid. (b) A zoomed-in image of particulates in one of the windows. (c) An amorphous as-deposited layer of the Ge-rich thin film is produced when moving away from the particulates. Figure S4: Optical constant data extracted from spectroscopic ellipsometry measurements and data fitting using Tauc–Lorentz optical oscillator. (a) Index of refraction for amorphous and (b) crystalline samples of Sb$_2$Te$_3$, GST225, and GST523 thin films. (c) The extinction coefficient for amorphous and (d) crystalline samples of Sb$_2$Te$_3$, GST225, GST523 thin films.

Author Contributions: D.T.Y. and A.J.T.V.D.R. contributed equally to the work. Conceptualization, B.J.K. and D.T.Y.; methodology and investigation, D.T.Y., A.J.T.V.D.R., O.A.E.K., J.M. and M.A.; formal analysis and visualization, D.T.Y., A.J.T.V.D.R. and O.A.E.K.; project administration and funding acquisition, B.J.K.; writing—original draft preparation, D.T.Y. and B.J.K.; writing—review and editing, A.J.T.V.D.R., J.M., M.B., M.A. and G.P.; supervision, B.J.K., M.B. and G.P. All authors have read and agreed to the published version of the manuscript.

Funding: This work has received funding from the European Union Horizon 2020 research and innovation program under Grant Agreement No. 824957 (BeforeHand: Boosting Performance of Phase Change Devices by Hetero- and Nanostructure Material Design).

Data Availability Statement: Data can be available upon request from the authors.

Conflicts of Interest: The authors declare no conflict of interest.

References

1. Wuttig, M.; Yamada, N. Phase-change materials for rewriteable data storage. *Nat. Mater.* **2007**, *6*, 824–832. [CrossRef] [PubMed]
2. Burr, G.W.; BrightSky, M.J.; Sebastian, A.; Cheng, H.Y.; Wu, J.Y.; Kim, S.; Sosa, N.E.; Papandreou, N.; Lung, H.L.; Pozidis, H.; et al. Recent Progress in Phase-Change Memory Technology. *IEEE J. Emerg. Sel. Top. Circuits Syst.* **2016**, *6*, 146–162. [CrossRef]
3. Ielmini, D.; Lavizzari, S.; Sharma, D.; Lacaita, A.L. Physical interpretation, modeling and impact on phase change memory (PCM) reliability of resistance drift due to chalcogenide structural relaxation. In Proceedings of the Technical Digest—International Electron Devices Meeting, IEDM, Washington, DC, USA, 10–12 December 2007; pp. 939–942.
4. Pirovano, A.; Lacaita, A.L.; Pellizzer, F.; Kostylev, S.A.; Benvenuti, A.; Bez, R. Low-field amorphous state resistance and threshold voltage drift in chalcogenide materials. *IEEE Trans. Electron. Devices* **2004**, *51*, 714–719. [CrossRef]
5. Zhang, W.; Ma, E. Unveiling the structural origin to control resistance drift in phase-change memory materials. *Mater. Today* **2020**, *41*, 156–176. [CrossRef]
6. Sebastian, A.; Krebs, D.; Le Gallo, M.; Pozidis, H.; Eleftheriou, E. A collective relaxation model for resistance drift in phase change memory cells. In Proceedings of the 2015 IEEE International Reliability Physics Symposium, Monterey, CA, USA, 19–23 April 2015; pp. MY51–MY56.
7. Sun, Z.; Zhou, J.; Blomqvist, A.; Johansson, B.; Ahuja, R. Formation of large voids in the amorphous phase-change memory Ge$_2$Sb$_2$Te$_5$ alloy. *Phys. Rev. Lett.* **2009**, *102*, 075504. [CrossRef]
8. Yang, T.Y.; Cho, J.Y.; Park, Y.J.; Joo, Y.C. Influence of dopants on atomic migration and void formation in molten Ge$_2$Sb$_2$Te$_5$ under high-amplitude electrical pulse. *Acta Mater.* **2012**, *60*, 2021–2030. [CrossRef]
9. Luong, M.A.; Agati, M.; Ratel Ramond, N.; Grisolia, J.; Le Friec, Y.; Benoit, D.; Claverie, A. On Some Unique Specificities of Ge-Rich GeSbTe Phase-Change Material Alloys for Nonvolatile Embedded-Memory Applications. *Phys. Status Solidi Rapid Res. Lett.* **2021**, *15*, 2000471. [CrossRef]
10. Zuliani, P.; Varesi, E.; Palumbo, E.; Borghi, M.; Tortorelli, I.; Erbetta, D.; Libera, G.D.; Pessina, N.; Gandolfo, A.; Prelini, C.; et al. Overcoming temperature limitations in phase change memories with optimized Ge$_x$Sb$_y$Te$_z$. *IEEE Trans. Electron. Devices* **2013**, *60*, 4020–4026. [CrossRef]
11. Cappelletti, P.; Annunziata, R.; Arnaud, F.; Disegni, F.; Maurelli, A.; Zuliani, P. Phase change memory for automotive grade embedded NVM applications. *J. Phys. D Appl. Phys.* **2020**, *53*, 193002. [CrossRef]
12. Zhou, X.; Xia, M.; Rao, F.; Wu, L.; Li, X.; Song, Z.; Feng, S.; Sun, H. Understanding phase-change behaviors of carbon-doped Ge$_2$Sb$_2$Te$_5$ for phase-change memory application. *ACS Appl. Mater. Interfaces* **2014**, *6*, 14207–14214. [CrossRef]
13. Choi, K.J.; Yoon, S.M.; Lee, N.Y.; Lee, S.Y.; Park, Y.S.; Yu, B.G.; Ryu, S.O. The effect of antimony-doping on Ge$_2$Sb$_2$Te$_5$, a phase change material. *Thin Solid Films* **2008**, *516*, 8810–8812. [CrossRef]
14. Privitera, S.; Rimini, E.; Bongiorno, C.; Pirovano, A.; Bez, R. Effects of dopants on the amorphous-to-fcc transition in Ge$_2$Sb$_2$Te$_5$ thin films. *Nucl. Instrum. Methods Phys. Res. Sect. B Beam Interact. Mater. Atoms* **2007**, *257*, 352–354. [CrossRef]
15. Luong, M.A.; Cherkashin, N.; Pécassou, B.; Sabbione, C.; Mazen, F.; Claverie, A. Effect of Nitrogen Doping on the Crystallization Kinetics of Ge$_2$Sb$_2$Te$_5$. *Nanomaterials* **2021**, *11*, 1729. [CrossRef] [PubMed]

16. Kim, K.H.; Chung, J.G.; Kyoung, Y.K.; Park, J.C.; Choi, S.J. Phase-change characteristics of nitrogen-doped Ge$_2$Sb$_2$Te$_5$ films during annealing process. *J. Mater. Sci. Mater. Electron.* **2011**, *22*, 52–55. [CrossRef]
17. Shelby, R.M.; Raoux, S. Crystallization dynamics of nitrogen-doped Ge$_2$Sb$_2$Te$_5$. *J. Appl. Phys.* **2009**, *105*, 104902. [CrossRef]
18. Privitera, S.; Rimini, E.; Zonca, R. Amorphous-to-crystal transition of nitrogen- and oxygen-doped Ge$_2$Sb$_2$Te$_5$ films studied by in situ resistance measurements. *Appl. Phys. Lett.* **2004**, *85*, 3044–3046. [CrossRef]
19. Yu, X.; Zhao, Y.; Li, C.; Hu, C.; Ma, L.; Fan, S.; Zhao, Y.; Min, N.; Tao, S.; Wang, Y. Improved multi-level data storage properties of germanium-antimony-tellurium films by nitrogen doping. *Scr. Mater.* **2017**, *141*, 120–124. [CrossRef]
20. Nolot, E.; Sabbione, C.; Pessoa, W.; Prazakova, L.; Navarro, G. Germanium, antimony, tellurium, their binary and ternary alloys and the impact of nitrogen: An X-ray photoelectron study. *Appl. Surf. Sci.* **2021**, *536*, 147703. [CrossRef]
21. Palumbo, E.; Zuliani, P.; Borghi, M.; Annunziata, R. Forming operation in Ge-rich Ge$_x$Sb$_y$Te$_z$ phase change memories. *Solid. State. Electron.* **2017**, *133*, 38–44. [CrossRef]
22. Redaelli, A.; Petroni, E.; Annunziata, R. Material and process engineering challenges in Ge-rich GST for embedded PCM. *Mater. Sci. Semicond. Process.* **2022**, *137*, 106184. [CrossRef]
23. Prazakova, L.; Nolot, E.; Martinez, E.; Fillot, F.; Rouchon, D.; Rochat, N.; Bernard, M.; Sabbione, C.; Morel, D.; Bernier, N.; et al. Temperature driven structural evolution of Ge-rich GeSbTe alloys and role of N-doping. *J. Appl. Phys.* **2020**, *128*, 215102. [CrossRef]
24. Thomas, O.; Mocuta, C.; Putero, M.; Richard, M.I.; Boivin, P.; Arnaud, F. Crystallization behavior of N-doped Ge-rich GST thin films and nanostructures: An in-situ synchrotron X-ray diffraction study. *Microelectron. Eng.* **2021**, *244–246*, 111573. [CrossRef]
25. Privitera, S.M.S.; Sousa, V.; Bongiorno, C.; Navarro, G.; Sabbione, C.; Carria, E.; Rimini, E. Atomic diffusion in laser irradiated Ge-rich GeSbTe thin films for phase change memory applications. *J. Phys. D Appl. Phys.* **2018**, *51*, 145103. [CrossRef]
26. Agati, M.; Vallet, M.; Joulié, S.; Benoit, D.; Claverie, A. Chemical phase segregation during the crystallization of Ge-rich GeSbTe alloys. *J. Mater. Chem. C* **2019**, *7*, 8720–8729. [CrossRef]
27. Abou El Kheir, O.; Bernasconi, M. High-throughput calculations on the decomposition reactions of off-stoichiometry GeSbTe alloys for embedded memories. *Nanomaterials* **2021**, *11*, 2382. [CrossRef]
28. Lee, Y.H.; Liao, P.J.; Hou, V.; Heh, D.; Nien, C.H.; Kuo, W.H.; Chen, G.T.; Yu, S.M.; Chen, Y.S.; Wu, J.Y.; et al. Composition segregation of Ge-rich GST and its effect on reliability. In Proceedings of the 2021 IEEE International Reliability Physics Symposium (IRPS), Monterey, CA, USA, 21–25 March 2021; IEEE: Piscataway, NJ, USA, 2021. [CrossRef]
29. Baldo, M.; Melnic, O.; Scuderi, M.; Nicotra, G.; Borghi, M.; Petroni, E.; Motta, A.; Zuliani, P.; Redaelli, A.; Ielmini, D. Modeling of virgin state and forming operation in embedded phase change memory (PCM). In Proceedings of the 2020 IEEE International Electron Devices Meeting (IEDM), Online, 12–18 December 2020; IEEE: Piscataway, NJ, USA, 2020. [CrossRef]
30. Goffart, L.; Pelissier, B.; Lefèvre, G.; Le-Friec, Y.; Vallée, C.; Navarro, G.; Reynard, J.P. Surface oxidation phenomena in Ge-rich GeSbTe alloys and N doping influence for Phase Change Memory applications. *Appl. Surf. Sci.* **2022**, *573*, 151514. [CrossRef]
31. Zuliani, P.; Palumbo, E.; Borghi, M.; Dalla Libera, G.; Annunziata, R. Engineering of chalcogenide materials for embedded applications of Phase Change Memory. *Solid State Electron.* **2015**, *111*, 27–31. [CrossRef]
32. Abou El Kheir, O.; Dragoni, D.; Bernasconi, M. Density functional simulations of decomposition pathways of Ge-rich GeSbTe alloys for phase change memories. *Phys. Rev. Mater.* **2021**, *5*, 095004. [CrossRef]
33. Kooi, B.J.; De Hosson, J.T.M. Electron diffraction and high-resolution transmission electron microscopy of the high temperature crystal structures of Ge$_x$Sb2Te3+x (x=1,2,3) phase change material. *J. Appl. Phys.* **2002**, *92*, 3584–3590. [CrossRef]
34. Van Pieterson, L.; Lankhorst, M.H.R.; Van Schijndel, M. Phase change memory technology. *J. Vac. Sci. Technol. B* **2005**, *97*, 223. [CrossRef]
35. Akola, J.; Jones, R.O. Structural phase transitions on the nanoscale: The crucial pattern in the phase-change materials Ge$_2$Sb$_2$Te$_5$ and GeTe. *Phys. Rev. B Condens. Matter Mater. Phys.* **2007**, *76*, 235201. [CrossRef]
36. Hegedüs, J.; Elliott, S.R. Microscopic origin of the fast crystallization ability of Ge-Sb-Te phase-change memorymaterials. *Nat. Mater.* **2008**, *7*, 399–405. [CrossRef] [PubMed]
37. Berthier, R.; Bernier, N.; Cooper, D.; Sabbione, C.; Hippert, F.; Noé, P. In situ observation of the impact of surface oxidation on the crystallization mechanism of GeTe phase-change thin films by scanning transmission electron microscopy. *J. Appl. Phys.* **2017**, *122*, 115304. [CrossRef]
38. Noé, P.; Sabbione, C.; Bernier, N.; Castellani, N.; Fillot, F.; Hippert, F. Impact of interfaces on scenario of crystallization of phase change materials. *Acta Mater.* **2016**, *110*, 142–148. [CrossRef]
39. Kooi, B.J.; Groot, W.M.G.; De Hosson, J.T.M. In situ transmission electron microscopy study of the crystallization of Ge$_2$Sb$_2$Te$_5$. *J. Appl. Phys.* **2004**, *95*, 924–932. [CrossRef]
40. Chen, B.; Lam Do, V.; Ten Brink, G.; Palasantzas, G.; Rudolf, P.; Kooi, B.J. Dynamics of GeSbTe phase-change nanoparticles deposited on graphene. *Nanotechnology* **2018**, *29*, 505706. [CrossRef]
41. Sun, L.; Zhou, Y.-X.; Wang, X.-D.; Chen, Y.-H.; Deringer, V.L.; Mazzarello, R.; Zhang, W. Ab initio molecular dynamics and materials design for embedded phase-change memory. *npj Comput. Mater.* **2021**, *7*, 29. [CrossRef]
42. Bordas, S.; Clavaguer-Mora, M.T.; Legendre, B.; Hancheng, C. Phase diagram of the ternary system Ge-Sb-Te. II. The subternary Ge-GeTe-Sb2Te3-Sb. *Thermochim. Acta* **1986**, *107*, 239–265. [CrossRef]

Article

Phase Change Ge-Rich Ge–Sb–Te/Sb$_2$Te$_3$ Core-Shell Nanowires by Metal Organic Chemical Vapor Deposition

Arun Kumar [1], Raimondo Cecchini [2], Claudia Wiemer [1], Valentina Mussi [3], Sara De Simone [3], Raffaella Calarco [3], Mario Scuderi [4], Giuseppe Nicotra [4] and Massimo Longo [3,*]

1. CNR—Institute for Microelectronics and Microsystems, Via C. Olivetti 2, 20864 Agrate Brianza, Italy; arun.kumar@mdm.imm.cnr.it (A.K.); claudia.wiemer@mdm.imm.cnr.it (C.W.)
2. CNR—Institute for Microelectronics and Microsystems, Via Gobetti 101, 40129 Bologna, Italy; cecchini@bo.imm.cnr.it
3. CNR—Institute for Microelectronics and Microsystems, Via del Fosso del Cavaliere 100, 00133 Rome, Italy; valentina.mussi@artov.imm.cnr.it (V.M.); sara.desimone@artov.imm.cnr.it (S.D.S.); raffaella.calarco@artov.imm.cnr.it (R.C.)
4. CNR—Institute for Microelectronics and Microsystems, Strada VIII 5, 95121 Catania, Italy; mario.scuderi@imm.cnr.it (M.S.); giuseppe.nicotra@imm.cnr.it (G.N.)
* Correspondence: massimo.longo@artov.imm.cnr.it

Abstract: Ge-rich Ge–Sb–Te compounds are attractive materials for future phase change memories due to their greater crystallization temperature as it provides a wide range of applications. Herein, we report the self-assembled Ge-rich Ge–Sb–Te/Sb$_2$Te$_3$ core-shell nanowires grown by metal-organic chemical vapor deposition. The core Ge-rich Ge–Sb–Te nanowires were self-assembled through the vapor–liquid–solid mechanism, catalyzed by Au nanoparticles on Si (100) and SiO$_2$/Si substrates; conformal overgrowth of the Sb$_2$Te$_3$ shell was subsequently performed at room temperature to realize the core-shell heterostructures. Both Ge-rich Ge–Sb–Te core and Ge-rich Ge–Sb–Te/Sb$_2$Te$_3$ core-shell nanowires were extensively characterized by means of scanning electron microscopy, high resolution transmission electron microscopy, X-ray diffraction, Raman microspectroscopy, and electron energy loss spectroscopy to analyze the surface morphology, crystalline structure, vibrational properties, and elemental composition.

Keywords: MOCVD; VLS; phase-change memory; nanowires; core-shell; Ge–Sb–Te; Ge–Sb–Te/Sb$_2$Te$_3$

1. Introduction

Chalcogenide materials have gained the interest of the microelectronic industry because of their fast data-storage speed, high endurance, reduced power consumption, high scalability, and cost effective production for the next generation of electronic phase change memories (PCM) [1–3]. The PCM data storage mechanism exploits the fast and reversible phase transitions of chalcogenide compounds, driven by electrical pulses, so that a resistivity contrast between the amorphous and crystalline phases is generated, to which digital information can be associated [4].

Over the last decades, several phase change compounds such as GeTe [5], Sb$_2$Te$_3$ [6,7], Ge$_2$Sb$_2$Te$_5$ (GST) [8,9], and In–Sb–Te [10] have been successfully synthesized. Indeed, GST is a typical phase-change compound exhibiting fast phase transitions between the amorphous and crystalline states. Such a structural modification involves the switching between the low and high electrical resistivity states associated with the two phases of the compound [11,12].

In order to obtain better device performances, the properties of the phase change materials can be optimized. GST exhibits a crystallization temperature (T_c) of about 150 °C with an activation energy of 2.5 eV [13], which is considered not enough to guarantee suitable data retention performances, especially for automotive applications. However, it has been found that the T_c can be varied by modifying the chemical composition of GST [14,15] or by introducing proper dopants such as carbon [16], nitrogen [17], titanium [18], and arsenic [19]. A large increase in T_c can be obtained by introducing a higher Ge content than that of GST [14]. This modification strongly improves the thermal stability of the PCM, leading to high data retention. Compared to GST and GeTe, Sb_2Te_3 exhibits a lower T_c, but faster reversible switching. Employing phase change alloys with different transition properties for the realization of memory cells based on heterostructures can improve the overall PCM performances as well as allow key studies on the role played by the heterointerface in the resistive switching mechanism. In fact, multilevel PCM cells based on $Ge_2Sb_2Te_5/Sb_2Te_3$ and $GeTe/Sb_2Te_3$ have been reported for the multilayer geometry [20,21]. An intrinsic advantage of cell down scaling is obtained by reducing the programming volume, hence lowering the energy needed for the phase transitions, therefore the power consumption. In this regard, nanowires (NWs) offer promising perspectives due to their low dimensionality and single crystalline structure [12,22–30]. However, up to now, the synthesis and the physical–chemical characterization of $Ge(Sb)Te/Sb_2Te_3$ core-shell NWs have not been reported, except for our recent study on $GeTe/Sb_2Te_3$ core shell NWs [31].

For this reason, we focused on the synthesis of a Ge-rich Ge–Sb–Te/Sb_2Te_3 NW based heterostructure, which could meet the industrial specifications from the point of view of the thermal stability of PCMs. The implementation of such a Ge-rich Ge–Sb–Te/Sb_2Te_3 heterostructure in final devices is still demanding, since it involves not only a precise NW positioning and contacting, but also a high control of the Ge–Sb–Te composition during the manufacturing process. Nevertheless, such a bottom-up route can avoid the most critical stage in the fabrication of the device, namely the pattering of the cells [32,33]. Moreover, the core-shell NWs are expected to be a useful term of comparison with the equivalent PCM cells formed by planar heterostructures, offering a direct device access with the additional benefit of a relatively simple PCM cell that can be obtained by metal contact definition on the NWs [13,15].

In this work, we report on the bottom-up, Au catalyzed, metal-organic chemical vapor deposition (MOCVD) growth of Ge-rich Ge–Sb–Te/Sb_2Te_3 core-shell NWs. An extensive analytical characterization of the obtained core-shell heterostructures was performed. The core-shell NWs were synthesized by self-assembling the Ge-rich Ge–Sb–Te core, which was subsequently coated with the Sb_2Te_3 shell, conformally overgrown. The growth study led to optimized core-shell NWs and the assessment of their morphological, compositional, and structural properties, in light of their functional analysis.

2. Experimental Methods

The NW growth was performed by an Aixtron AIX200/4 MOCVD tool (Aixtron SE, Herzogenrath, Germany), employing the vapor–liquid–solid (VLS) mechanism catalyzed by different Au nanoparticles (NPs) with average sizes of 10 nm, 20 nm, 30 nm, and 50 nm. The Au NPs, in a colloidal solution from Ted Pella© (Redding, CA, USA), were dispersed on Si (100) and SiO_2/Si substrates via a simple drop-casting method. The native oxide on the Si (100) substrates was removed by immersion in a HF 5% solution prior to dispersion of the Au NPs. The employed metal-organic sources were of electronic-grade tetrakisdimethylamino germanium ($Ge[N(CH_3)_2]_4$, TDMAGe), antimonytrichloride ($SbCl_3$), diisopropyltelluride (($C_3H_7)_2$Te, DiPTe), and bis(trimethylsilyl) telluride ($Te(SiMe_3)_2$, DSMTe), carried to the MOCVD reactor by an ultra-pure N_2 process gas. The Ge-rich Ge–Sb–Te core NWs were obtained by employing the TDMAGe, $SbCl_3$, and DiPTe precursors with the reactor pressure (P), reactor temperature (T), and growth duration (t) ranging in the values of 50–300 mbar, 300–400 °C, and 60–120 min, respectively. The Sb_2Te_3 shell overgrowth on the Ge-rich Ge–Sb–Te core NWs was performed at room temperature, P = 15 mbar, t = 90 min, $SbCl_3$ partial pressure = 2.23×10^{-4} mbar, and DSMTe partial pressure = 3.25×10^{-4} mbar.

The surface morphology of the obtained NWs was investigated by a ZeissR©Supra40 field-emission scanning electron microscope (FE-SEM) (Cark Zeiss, Oberkochen, Germany) in plan as well as in cross-section mode. X-ray diffraction (XRD) experiments with an

ItalStructures HRD3000 diffractometer system (Ital Structures Sas, Riva de Garda, Italy) were performed to pattern the average crystal structure of the obtained NWs. The obtained XRD patterns were analyzed using a best-fit process based on the Rietveld method.

High resolution transmission electron microscopy (HR-TEM) was employed to investigate the local microstructure, elemental composition, and the growth direction. For the TEM characterization, the as-synthesized NWs from the substrate were transferred directly on the TEM grid by mechanical rubbing. TEM characterization was carried out using a Cs-probe-corrected TEM JEOL ARM200CF microscope (JEOL Ltd., Akishima, Japan), operated at 200 keV and equipped with a GIF Quantum ER energy filter from GATAN (Gatan Inc., Pleasanton, CA, USA) for electron energy loss spectroscopy (EELS). Micro Raman characterizations were carried out by means of a Thermo Scientific DXR2xi Raman Imaging Microscope (Thermo Fischer Scientific, Waltham, MA, USA), coupled with a 532 nm excitation laser and a 50× objective. The obtained spectra were recorded after 40 accumulations of 20 ms acquisitions.

3. Results and Discussion

Initially, we investigated the growth of the Ge-rich Ge–Sb–Te core NWs. In this regard, we systematically explored the deposition on both Si (100) and SiO$_2$/Si substrates with different Au NP sizes and it has to be noted that, depending on the precursor choice and deposition condition, different NW morphology, density, and compositions could be obtained from pure Ge to Sb$_2$Te$_3$ NWs (see Supplementary Materials, Figures S1–S6). Table S1, summarizes the growth parameters related to SEM images displayed in Figure S1, Supplementary Materials. In fact, at T = 400 °C, P = 50 mbar, and t = 60 min, the NWs with the best morphological characteristics in terms of length and density were achieved. Starting from this point, we obtained Ge-rich Ge–Sb–Te core NWs, with TDMAGe, SbCl$_3$, and DiPTe bubblers' partial pressures set to 3.35×10^{-3} mbar, 5.12×10^{-5} mbar, and 8.58×10^{-3} mbar, respectively. The plan view SEM images on Ge-rich Ge–Sb–Te core NWs grown with 10 nm, 20 nm, 30 nm, and 50 nm Au NPs on SiO$_2$/Si, and Si (100) substrates demonstrated the formation of a good density of Ge-rich Ge–Sb–Te NWs via the VLS mechanism (see Supplementary Materials, Figure S5). The obtained NWs on SiO$_2$/Si were found to have an average length and diameter distribution up to 1.40 µm and 100 nm, respectively (see Supplementary Materials, Figure S7).

Figure 1a,b shows the planar and cross-sectional SEM images for Ge-rich Ge–Sb–Te core, and Ge-rich Ge–Sb–Te/Sb$_2$Te$_3$ core-shell NWs catalyzed by Au NPs of 50 nm size, respectively. Interestingly, the SiO$_2$/Si surface exhibited a decent density of NWs all over the substrate (Figure 1a). The VLS growth mechanism in the obtained NWs could be confirmed by the presence of the catalyst Au NPs at the tips of the NWs.

In the second step of the core-shell fabrication, the Ge-rich Ge–Sb–Te NWs were coated with a Sb$_2$Te$_3$ deposition under the MOCVD growth conditions: T = room temperature, P = 15 mbar, t = 90 min, leading to the conformal growth of a uniform and continuous layer all over the core NWs, irrespectively of the different NP sizes (see Figure 1c,d and Supplementary Materials, Figure S6e–h). For the shell, it was necessary to lower the deposition temperature down to room temperature to obtain the conformal NW coating, although with a granular morphology. The obtained Ge-rich Ge–Sb–Te/Sb$_2$Te$_3$ core-shell NWs exhibited a conformal overgrowth of about 30 nm, with approximate inclusive diameters ranging from 90 nm to 130 nm. The obtained results showed that, with the appropriate growth condition selection, NWs with relatively high density and reproducibility can be achieved.

Figure 2 shows a set of XRD patterns obtained for the Ge-rich Ge–Sb–Te core, and Ge-rich Ge–Sb–Te/Sb$_2$Te$_3$ core-shell NWs on a SiO$_2$/Si substrate. The Ge-rich Ge–Sb–Te core NWs exhibited broad diffraction peaks centered at the 2θ values likely for the face centered cubic (FCC), Ge$_2$Sb$_2$Te$_5$ phases, along with the presence of Au NP diffraction peaks (Figure 2a). No presence of amorphous Ge was detected in the NWs or additional structures, as investigated by SEM. Therefore, the Ge-rich Ge–Sb–Te is crystallized in the cubic structure, with a lattice parameter value similar to that of cubic Ge$_2$Sb$_2$Te$_5$. The additional Ge is likely to be present as an interstitial element in the lattice, causing the spread of the lattice parameter value, which is detected as the enhanced broadness of the diffracted maxima. It should be noted that the cubic GST phase could be simply transformed into the hexagonal structure with a minor atomic rearrangement during the transition process [34]. However, in Figure 2b, for Ge-rich Ge–Sb–Te/Sb_2Te_3 core-shell NWs, the GST peak became less intense and narrower, while the diffraction peaks from the Sb_2Te_3 phase were clearly obtained. This suggested that no structural disordering occurred after the shell deposition.

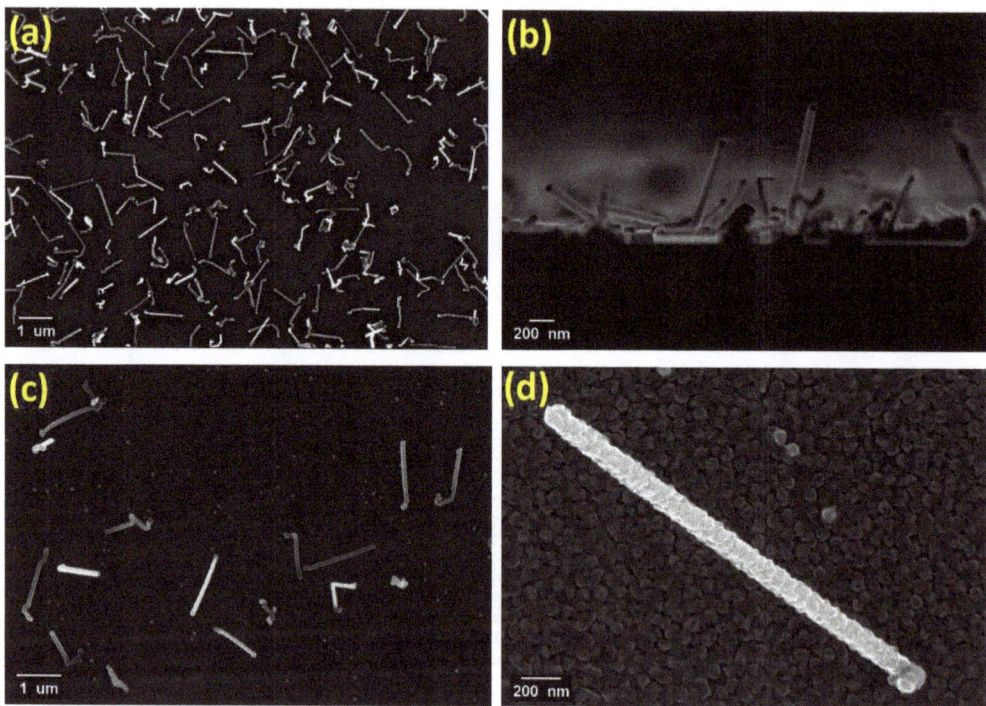

Figure 1. SEM images in (**a**) plan, (**b**) cross-section view of Ge-rich Ge–Sb–Te core NWs; and (**c**,**d**) plan view of Ge-rich Ge–Sb–Te/Sb_2Te_3 core-shell NWs on a SiO_2/Si substrate, with different magnifications, respectively.

Micro-Raman mapping was carried out on the deposited Ge-rich Ge–Sb–Te core, and Ge-rich Ge–Sb–Te/Sb_2Te_3 core-shell NWs to analyze the vibrational modes, obtaining similar spectral results on both SiO_2/Si and Si (100) substrates. Figure 3a shows the dark field optical image of Ge-rich Ge–Sb–Te NWs identified with a Raman microscope on the Si (100) substrate with Au NPs 50 nm in size. The Raman map corresponding to the area indicated by the red rectangle is presented in Figure 3b and was carried out with a 0.5 μm step size. In Figure 3c, the mean spectrum calculated over the wire (green area of the map) can be clearly distinguished from that obtained on the background (blue area of the map). The NW Raman spectrum mainly showed two peaks at about 127 and 142 cm^{-1}, and could be attributed to the FCC-$Ge_2Sb_2Te_5$ phase [35]. The obtained results are in good agreement with the XRD results, thus the structural and vibrational analysis confirmed the FCC phase of the obtained Ge-rich Ge–Sb–Te NWs.

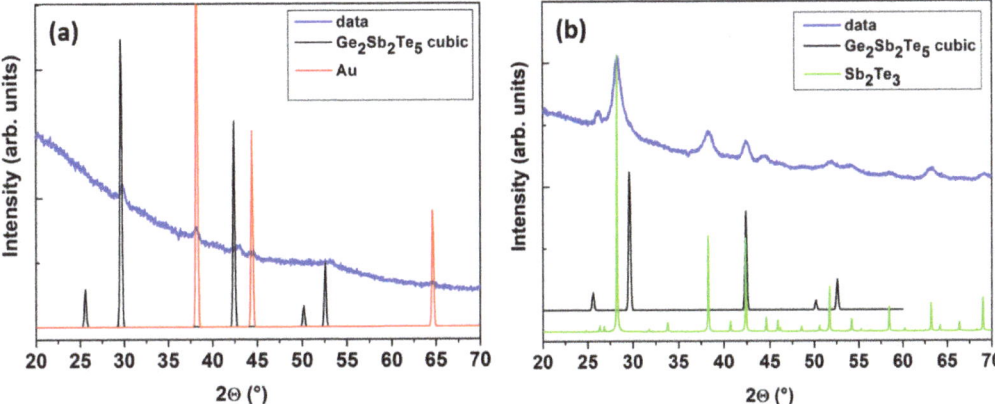

Figure 2. XRD pattern on (**a**) Ge-rich Ge–Sb–Te core NWs, and (**b**) Ge-rich Ge–Sb–Te/Sb$_2$Te$_3$ core-shell NW, with 50 nm Au NPs on the SiO$_2$/Si substrate, respectively.

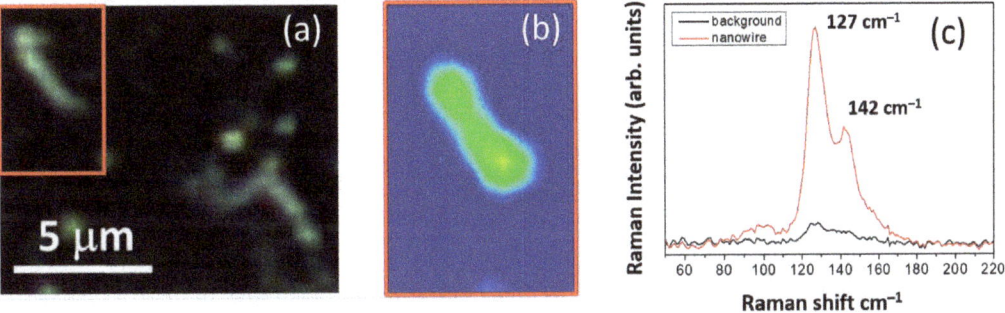

Figure 3. (**a**) Optical image of a Ge-rich Ge–Sb–Te NW identified by Raman microscope on a Si (100) substrate with 50 nm of Au NPs; (**b**) Raman map corresponding to the area indicated by the red rectangle in the optical image; (**c**) Mean Raman spectrum calculated over the wire, green area of the map, compared with that obtained on the background, blue area of the map.

Figure 4 shows the micro Raman image obtained on the Ge-rich Ge–Sb–Te/Sb$_2$Te$_3$ core-shell NWs on a Si (100) substrate with Au NPs of 50 nm size. A single NW was optically localized and selected by Figure 4a and then spectrally mapped (Figure 4b). Figure 4c reports the mean spectrum calculated over the NW (red area of the map), which shows three dominant peaks at about 69 cm^{-1}, 112 cm^{-1}, and 167 cm^{-1}, associated with the A^1_{1g} (LO), E^2_g (TO), and A^2_{1g} (LO) modes of Sb$_2$Te$_3$, respectively [36–38], in addition to those already identified for Ge-rich Ge–Sb–Te at about 127, and 142 cm^{-1}.

The HRTEM investigation demonstrated that the obtained NWs have a well-defined core-shell heterostructure, where the core consists of single crystalline Ge-rich Ge–Sb–Te, surrounded by a polycrystalline Sb$_2$Te$_3$ shell. Figure 5a reports a bright-field TEM micrograph of a Ge-rich Ge–Sb–Te core NW with a size of 90 nm. The image shows that the Ge-rich Ge–Sb–Te NW has a smooth surface with a clearly evident Au catalyst particle at the tip, demonstrating the occurrence of the VLS mechanism during the growth process. The selected area diffraction (SAED) pattern showed regular spot patterns, which further confirmed the perfect single crystalline nature of the NWs. This means that the NWs are made by a single FCC Ge–Sb–Te crystal, as indicated by the respective SAED pattern taken on the trunk (Figure 5b). The EELS analysis (not shown here) confirmed an estimated composition of 35% Ge:10% Sb:55% Te throughout the core NW (see Supplementary Materials,

Figure S8), corresponding to the atomic ratio of Ge:Sb:Te = 3:1:5. The obtained composition had a higher Ge concentration with respect to the stoichiometric GST.

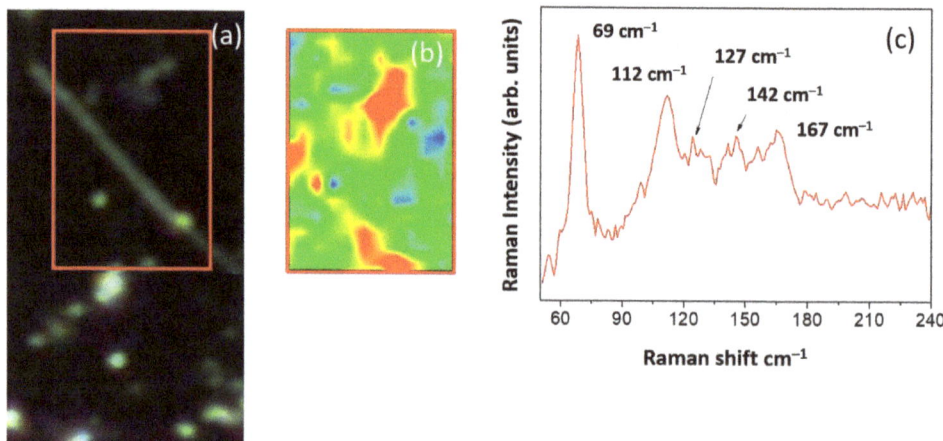

Figure 4. (**a**) Optical image of a Ge-rich Ge–Sb–Te/Sb$_2$Te$_3$ core-shell NW on a Si (100) substrate with 50 nm of Au NPs identified by the Raman microscope; (**b**) Raman map corresponding to the area indicated by the red rectangle in the optical image; (**c**) Mean Raman spectrum calculated over the wire, red area of the map.

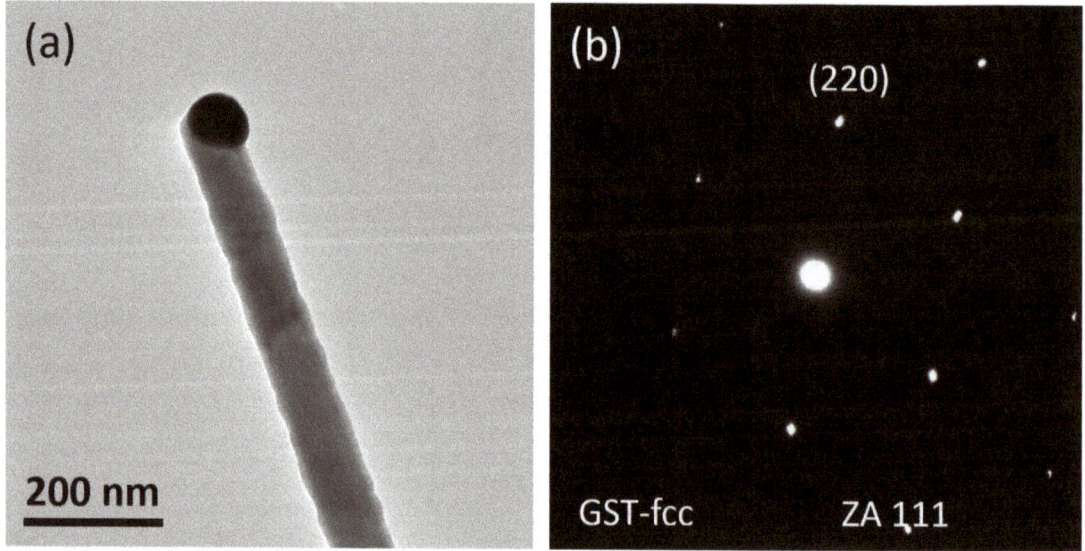

Figure 5. Bright-field TEM micrograph of a Ge-rich Ge–Sb–Te core NW on SiO$_2$/Si substrate (**a**) and the relative SAED patterns (**b**), showing a FCC phase.

Furthermore, the TEM images of Ge-rich Ge–Sb–Te/Sb$_2$Te$_3$ core-shell NWs indicated a uniform and homogenous conformal overgrowth around the core NWs. In fact, one can clearly observe an interface between the core and the shell region of the NWs, as displayed in Figure 6a, where a bright-field TEM micrograph of a Ge-rich Ge–Sb–Te/Sb$_2$Te$_3$ core-shell NW with a core size of about 60 nm and a Sb$_2$Te$_3$ conformal shell about 30 nm thick, is reported. The respective SAED pattern in Figure 6b shows the superposition of a regular diffraction pattern that originates from the monocrystalline core and other dispersed spots set out in the rings around the central spot that originates from the polycrystalline shell. EELS investigation of the polycrystalline shell deposited over the core NW confirmed the existence of only Sb and Te elements, with a stoichiometric ratio of a 2:3, respectively (see Supplementary Materials, Figure S9). In conclusion, the Ge-rich Ge–Sb–Te/Sb_2Te_3 core-shell NWs were found without any defects such as stacking faults and dislocations; the crystal structures observed by TEM all over the core and core-shell NW heterostructures were in good agreement with those observed by the XRD analysis.

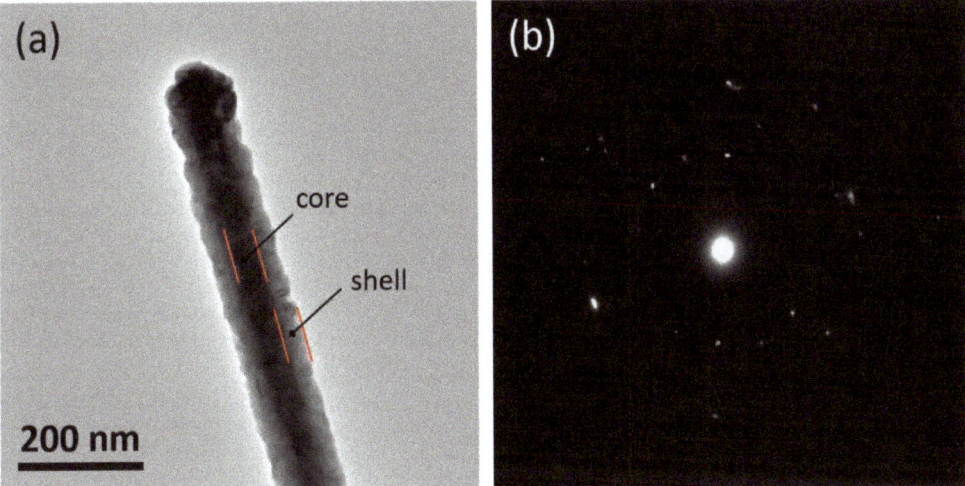

Figure 6. Bright-field TEM micrograph of a Ge-rich Ge–Sb–Te/Sb_2Te_3 core-shell NW on the SiO_2/Si substrate (**a**) and relative SAED patterns (**b**) showing the polycrystalline diffraction pattern from the shell.

The above discussed results validate the capability of the presented method in achieving both high quality Ge-rich Ge–Sb–Te core NWs and a uniform Sb_2Te_3 shell, conformally deposited at room temperature. This could be achieved while preserving a physical–chemical sharp interface, with minimum elemental inter-diffusion between the two alloys, favored by the room temperature deposition of the Sb_2Te_3 shell. These findings also indicate that a reaction between the residual Ge–Sb–Te precursor and the Sb_2Te_3 precursor could be avoided and no undesired alloys were formed.

4. Conclusions

We reported the optimized MOCVD synthesis of crystalline ternary Ge-rich Ge–Sb–Te (35% Ge, 10% Sb, 55% Te) core nanowires, catalyzed by the VLS mechanism, with a good density. The as-synthesized Ge-rich Ge–Sb–Te nanowires were single crystals in the FCC phase, with an average length up to 1.40 µm, diameter less than 60 nm, and higher in Ge concentration with respect to the stoichiometric $Ge_2Sb_2Te_5$. The conformal overgrowth of polycrystalline Sb_2Te_3 was achieved at room temperature on the previously grown core Ge-rich Ge–Sb–Te nanowires. No structural disordering and core-shell inter-diffusion was evidenced. The obtained chalcogenide core-shell nanowires will be explored for the phase transition behavior, where nanoscale phenomenon and interface effects can be investigated for future embedded PCM cells and automotive applications.

Supplementary Materials: The following are available online at https://www.mdpi.com/article/10.3390/nano11123358/s1, Figure S1: Plan view SEM images on MOCVD growth of Ge–Sb–Te alloy with 10 nm Au NP on SiO_2/Si substrates, Figure S2: (a) STEM image of as grown Ge–Sb–Te with 10 nm Au NP on SiO_2/Si substrates at T = 400 °C, P = 50 mbar, t = 120 min, TDMAGe partial pressure = 2.94 × 10^{-3} mbar, SbCl$_3$ partial pressure = 2.42 × 10^{-3} mbar, DSMTe partial pressure = 3.52 × 10^{-3} mbar, (b) the composition of the NWs resulted to be mainly that of Ge, as indicated by the EDX spectrum, Figure S3: (a) STEM-EDX images showing the presence of NWs with 10 nm Au NP on SiO$_2$/Si substrates at T = 380 °C, P = 300 mbar, t = 180 min, TDMAGe partial pressure = 4.77 × 10^{-3} mbar, SbCl$_3$ partial pressure = 1.38 × 10^{-3} mbar, DSMTe partial pressure = 5.37 × 10^{-3} mbar; (b) the composition of the NWs resulted to be mainly that of Sb$_2$Te$_3$, as indicated by the EDX spectrum, Figure S4: XRD analysis of NWs with 10 nm Au NP on SiO$_2$/Si substrates at T = 380 °C, P = 300 mbar, t = 180 min, TDMAGe partial pressure = 4.77 × 10^{-3} mbar, SbCl$_3$ partial pressure = 1.38 × 10^{-3} mbar, DSMTe partial pressure = 5.37 × 10^{-3} mbar, Figure S5: Plan view SEM images on MOCVD growth of Ge–Sb–Te nanowires with 50 nm Au NP on Si(100) and SiO$_2$/Si substrates at T = 400 °C, P = 50 mbar, t = 60 min, TDMAGe partial pressure = 3.35 × 10^{-3} mbar, (a,b) SbCl$_3$ partial pressure = 2.07 × 10^{-3} mbar, DiPTe partial pressure = 7.07 × 10^{-3}, (c,d) SbCl$_3$ partial pressure = 1.04 × 10^{-3} mbar, DiPTe partial pressure = 7.07 × 10^{-3} mbar, (e,f) SbCl$_3$ partial pressure = 3.45 × 10^{-4} mbar, DiPTe partial pressure = 7.07 × 10^{-3} mbar, (g,h) SbCl$_3$ partial pressure = 1.73 × 10^{-4} mbar, DiPTe partial pressure = 8.58 × 10^{-3} mbar, (i,j) SbCl$_3$ partial pressure = 5.18 × 10^{-5} mbar, DiPTe partial pressure = 8.58 × 10^{-3} mbar, respectively, Figure S6: Plan view SEM images on MOCVD growth of (a–d) Ge-rich Ge–Sb–Te core nanowires, and (e–h) Ge-rich Ge–Sb–Te/Sb$_2$Te$_3$ core-shell NWs with 10, 20, 30 and 50 nm Au NPs on SiO$_2$/Si substrate, respectively, Figure S7: Statistical plot of the average length and diameter distribution of Ge-rich Ge–Sb–Te core nanowires on SiO$_2$/Si with 10, 20, 30 and 50 nm Au NPs, Figure S8: STEM image of a Ge-rich Ge–Sb–Te core nanowire portion (a) and its corresponding EELS spectra for Ge L-edge (b) and Sb–Te M-edge (c), Figure S9: STEM image of a Ge-rich Ge–Sb–Te/Sb$_2$Te$_3$ core-shell nanowire portion (a) and its corresponding EELS spectrum (b), Table S1: Growth parameters corresponding to Figure S1.

Author Contributions: Conceptualization, M.L.; Funding acquisition, G.N. and M.L.; Investigation, A.K., R.C. (Raimondo Cecchini), C.W., V.M., S.D.S., R.C. (Raffaella Calarco) and M.S.; Supervision, M.L.; Writing—original draft, A.K.; Writing—review & editing, R.C. (Raimondo Cecchini), C.W., V.M., S.D.S., R.C. (Raffaella Calarco), M.S., G.N. and M.L. All authors have read and agreed to the published version of the manuscript.

Funding: This project has received funding from the European Union's Horizon 2020 research and innovation program under Grant Agreement No. 824957 ("BeforeHand: Boosting Performance of Phase Change Devices by Hetero- and Nanostructure Material Design"). TEM measurements were performed at BeyondNano CNR-IMM, which is supported by the Italian Ministry of Education and Research (MIUR) under project Beyond-Nano (PON a3_00363). The STEM received funding from the European Union's Horizon 2020 research and innovation program under grant agreement no. 823717–ESTEEM3.

Institutional Review Board Statement: Not applicable.

Informed Consent Statement: Not applicable.

Data Availability Statement: The data that support the findings of this study are available from the corresponding author upon reasonable request.

Conflicts of Interest: The authors declare no conflict of interest.

References

1. Raoux, S.; Wełnic, W.; Ielmini, D. Phase Change Materials and Their Application to Nonvolatile Memories. *Chem. Rev.* **2009**, *110*, 240–267. [CrossRef]
2. Zhang, W.; Mazzarello, R.; Wuttig, M.; Ma, E. Designing crystallization in phase-change materials for universal memory and neuro-inspired computing. *Nat. Rev. Mater.* **2019**, *4*, 150–168. [CrossRef]
3. Le Gallo, M.; Sebastian, A. An overview of phase-change memory device physics. *J. Phys. D. Appl. Phys.* **2020**, *53*, 213002. [CrossRef]
4. Raoux, S.; Xiong, F.; Wuttig, M.; Pop, E. Phase change materials and phase change memory. *MRS Bull.* **2014**, *39*, 703–710. [CrossRef]
5. Singh, K.; Kumari, S.; Singh, H.; Bala, N.; Singh, P.; Kumar, A.; Thakur, A. A review on GeTe thin film-based phase-change materials. *Appl. Nanosci.* **2021**, *2021*, 1–16. [CrossRef]
6. Jacobs-Gedrim, R.B.; Murphy, M.T.; Yang, F.; Jain, N.; Shanmugam, M.; Song, E.S.; Kandel, Y.; Hesamaddin, P.; Yu, H.Y.; Anantram, M.P.; et al. Reversible phase-change behavior in two-dimensional antimony telluride (Sb$_2$Te$_3$) nanosheets. *Appl. Phys. Lett.* **2018**, *112*, 133101. [CrossRef]

7. Liu, G.Y.; Wu, L.C.; Song, Z.T.; Rao, F.; Song, S.N.; Cheng, Y. Stability of Sb2Te Crystalline Films for Phase Change Memory. *Mater. Sci. Forum* **2017**, *898*, 1829–1833. [CrossRef]
8. Guo, P.; Sarangan, A.M.; Agha, I. A Review of Germanium-Antimony-Telluride Phase Change Materials for Non-Volatile Memories and Optical Modulators. *Appl. Sci.* **2019**, *9*, 530. [CrossRef]
9. Loke, D.; Lee, T.H.; Wang, W.J.; Shi, L.P.; Zhao, R.; Yeo, Y.C.; Chong, T.C.; Elliott, S.R. Breaking the Speed Limits of Phase-Change Memory. *Science* **2012**, *336*, 1566–1569. [CrossRef]
10. Fallica, R.; Stoycheva, T.; Wiemer, C.; Longo, M. Structural and electrical analysis of In–Sb–Te-based PCM cells. *Phys. Status Solidi—Rapid Res. Lett.* **2013**, *7*, 1009–1013. [CrossRef]
11. Lee, S.H.; Yeonwoong, J.; Agarwal, R. Size-Dependent Surface-Induced heterogeneous nucleation driven Phase-Change in Ge 2Sb 2Te 5 nanowires. *Nano Lett.* **2008**, *8*, 3303–3309. [CrossRef]
12. Jung, C.S.; Kim, H.S.; Im, H.S.; Seo, Y.S.; Park, K.; Back, S.H.; Cho, Y.J.; Kim, C.H.; Park, J.; Ahn, J.P. Polymorphism of GeSbTe superlattice nanowires. *Nano Lett.* **2013**, *13*, 543–549. [CrossRef] [PubMed]
13. Pellizzer, F.; Benvenuti, A.; Gleixner, B.; Kim, Y.; Johnson, B.; Magistretti, M.; Marangon, T.; Pirovano, A.; Bez, R.; Atwood, G. A 90nm phase change memory technology for stand-alone non-volatile memory applications. In Proceedings of the 2006 Symposium on VLSI Technology, Honolulu, HI, USA, 13–15 June 2006; pp. 122–123. [CrossRef]
14. Zuliani, P.; Varesi, E.; Palumbo, E.; Borghi, M.; Tortorelli, I.; Erbetta, D.; Libera, G.D.; Pessina, N.; Gandolfo, A.; Prelini, C.; et al. Overcoming temperature limitations in phase change memories with optimized Gex Sby Tez. *IEEE Trans. Electron. Devices* **2013**, *60*, 4020–4026. [CrossRef]
15. Cheng, H.Y.; Hsu, T.H.; Raoux, S.; Wu, J.Y.; Du, P.Y.; Breitwisch, M.; Zhu, Y.; Lai, E.K.; Joseph, E.; Mittal, S.; et al. A high performance phase change memory with fast switching speed and high temperature retention by engineering the Ge xSb yTe z phase change material. In Proceedings of the Technical Digest—International Electron Devices Meeting, IEDM, Washington, DC, USA, 5–7 December 2011.
16. Li, T.; Shen, J.; Wu, L.; Song, Z.; Lv, S.; Cai, D.; Zhang, S.; Guo, T.; Song, S.; Zhu, M. Atomic-Scale Observation of Carbon Distribution in High-Performance Carbon-Doped $Ge_2Sb_2Te_5$ and Its Influence on Crystallization Behavior. *J. Phys. Chem. C* **2019**, *123*, 13377–13384. [CrossRef]
17. Luong, M.A.; Cherkashin, N.; Pecassou, B.; Sabbione, C.; Mazen, F.; Claverie, A. Effect of Nitrogen Doping on the Crystallization Kinetics of $Ge_2Sb_2Te_5$. *Nanomaterials* **2021**, *11*, 1729. [CrossRef] [PubMed]
18. Wei, S.J.; Zhu, H.F.; Chen, K.; Xu, D.; Li, J.; Gan, F.X.; Zhang, X.; Xia, Y.J.; Li, G.H. Phase change behavior in titanium-doped $Ge_2Sb_2Te_5$ films. *Appl. Phys. Lett.* **2011**, *98*, 231910. [CrossRef]
19. Madhavan, V.E.; Carignano, M.; Kachmar, A.; Sangunni, K.S. Crystallization properties of arsenic doped GST alloys. *Sci. Rep.* **2019**, *9*, 1–10. [CrossRef] [PubMed]
20. Rao, F.; Song, Z.; Zhong, M.; Wu, L.; Feng, G.; Liu, B.; Feng, S.; Chen, B. Multilevel data storage characteristics of phase change memory cell with doublelayer chalcogenide films ($Ge_2Sb_2Te_5$ and Sb2Te3). *Jpn. J. Appl. Phys. Part 2 Lett.* **2007**, *46*, L25. [CrossRef]
21. Chong, T.C.; Shi, L.P.; Wei, X.Q.; Zhao, R.; Lee, H.K.; Yang, P.; Du, A.Y. Crystalline amorphous semiconductor superlattice. *Phys. Rev. Lett.* **2008**, *100*, 136101. [CrossRef] [PubMed]
22. Yu, D.; Wu, J.; Gu, Q.; Park, H. Germanium telluride nanowires and nanohelices with memory-switching behavior. *J. Am. Chem. Soc.* **2006**, *128*, 8148–8149. [CrossRef]
23. Longo, M.; Wiemer, C.; Salicio, O.; Fanciulli, M.; Lazzarini, L.; Rotunno, E. Au-catalyzed self assembly of GeTe nanowires by MOCVD. *J. Cryst. Growth* **2011**, *315*, 152–156. [CrossRef]
24. Nukala, P.; Lin, C.C.; Composto, R.; Agarwal, R. Ultralow-power switching via defect engineering in germanium telluride phase-change memory devices. *Nat. Commun.* **2016**, *7*, 10482. [CrossRef] [PubMed]
25. Meister, S.; Peng, H.; McIlwrath, K.; Jarausch, K.; Zhang, X.F.; Cui, Y. Synthesis and characterization of phase-change nanowires. *Nano Lett.* **2006**, *6*, 1514–1517. [CrossRef] [PubMed]
26. Longo, M.; Stoycheva, T.; Fallica, R.; Wiemer, C.; Lazzarini, L.; Rotunno, E. Au-catalyzed synthesis and characterisation of phase change Ge-doped Sb-Te nanowires by MOCVD. *J. Cryst. Growth* **2013**, *370*, 323–327. [CrossRef]
27. Jung, Y.; Lee, S.H.; Ko, D.K.; Agarwal, R. Synthesis and characterization of $Ge_2Sb_2Te_5$ nanowires with memory switching effect. *J. Am. Chem. Soc.* **2006**, *128*, 14026–14027. [CrossRef]
28. Cecchini, R.; Selmo, S.; Wiemer, C.; Fanciulli, M.; Rotunno, E.; Lazzarini, L.; Rigato, M.; Pogany, D.; Lugstein, A.; Longo, M. In-doped Sb nanowires grown by MOCVD for high speed phase change memories. *Micro Nano Eng.* **2019**, *2*, 117–121. [CrossRef]
29. Sun, X.; Yu, B.; Ng, G.; Nguyen, T.D.; Meyyappan, M. III-VI compound semiconductor indium selenide (In_2Se_3) nanowires: Synthesis and characterization. *Appl. Phys. Lett.* **2006**, *89*, 233121. [CrossRef]
30. Selmo, S.; Cecchini, R.; Cecchi, S.; Wiemer, C.; Fanciulli, M.; Rotunno, E.; Lazzarini, L.; Rigato, M.; Pogany, D.; Lugstein, A.; et al. Low power phase change memory switching of ultra-thin In3Sb1Te2 nanowires. *Appl. Phys. Lett.* **2016**, *109*, 213103. [CrossRef]
31. Kumar, A.; Cecchini, R.; Wiemer, C.; Mussi, V.; De Simone, S.; Calarco, R.; Scuderi, M.; Nicotra, G.; Longo, M. MOCVD Growth of GeTe/Sb_2Te_3 Core–Shell Nanowires. *Coatings* **2021**, *11*, 718. [CrossRef]
32. Canvel, Y.; Lagrasta, S.; Boixaderas, C.; Barnola, S.; Mazel, Y.; Martinez, E. Study of Ge-rich GeSbTe etching process with different halogen plasmas. *J. Vac. Sci. Technol. A* **2019**, *37*, 031302. [CrossRef]
33. Canvel, Y.; Lagrasta, S.; Boixaderas, C.; Barnola, S.; Mazel, Y.; Dabertrand, K.; Martinez, E. Modification of Ge-rich GeSbTe surface during the patterning process of phase-change memories. *Microelectron. Eng.* **2020**, *221*, 111183. [CrossRef]

34. Rotunno, E.; Lazzarini, L.; Longo, M.; Grillo, V. Crystal structure assessment of Ge-Sb-Te phase change nanowires. *Nanoscale* **2013**, *5*, 1557–1563. [CrossRef] [PubMed]
35. Pitchappa, P.; Kumar, A.; Prakash, S.; Jani, H.; Venkatesan, T.; Singh, R. Chalcogenide Phase Change Material for Active Terahertz Photonics. *Adv. Mater.* **2019**, *31*, 1808157. [CrossRef] [PubMed]
36. Sosso, G.C.; Caravati, S.; Bernasconi, M. Vibrational properties of crystalline Sb2Te3 from first principles. *J. Phys. Condens. Matter* **2009**, *21*, 095410. [CrossRef] [PubMed]
37. Shahil, K.M.F.; Hossain, M.Z.; Goyal, V.; Balandin, A.A. Micro-Raman spectroscopy of mechanically exfoliated few-quintuple layers of Bi_2Te_3, Bi_2Se_3, and Sb_2Te_3 materials. *J. Appl. Phys.* **2012**, *111*, 054305. [CrossRef]
38. Cecchi, S.; Dragoni, D.; Kriegner, D.; Tisbi, E.; Zallo, E.; Arciprete, F.; Holý, V.; Bernasconi, M.; Calarco, R. Interplay between Structural and Thermoelectric Properties in Epitaxial $Sb_{2+x}Te_3$ Alloys. *Adv. Funct. Mater.* **2019**, *29*, 1805184. [CrossRef]

Article

High-Throughput Calculations on the Decomposition Reactions of Off-Stoichiometry GeSbTe Alloys for Embedded Memories

Omar Abou El Kheir and Marco Bernasconi *

Dipartimento di Scienza dei Materiali, Università di Milano-Bicocca, Via R. Cozzi 55, I-20125 Milano, Italy; o.abouelkheir@campus.unimib.it
* Correspondence: marco.bernasconi@unimib.it

Abstract: Chalcogenide GeSbTe (GST) alloys are exploited as phase change materials in a variety of applications ranging from electronic non-volatile memories to neuromorphic and photonic devices. In most applications, the prototypical $Ge_2Sb_2Te_5$ compound along the $GeTe$-Sb_2Te_3 pseudobinary line is used. Ge-rich GST alloys, off the pseudobinary tie-line with a crystallization temperature higher than that of $Ge_2Sb_2Te_5$, are currently explored for embedded phase-change memories of interest for automotive applications. During crystallization, Ge-rich GST alloys undergo a phase separation into pure Ge and less Ge-rich alloys. The detailed mechanisms underlying this transformation are, however, largely unknown. In this work, we performed high-throughput calculations based on Density Functional Theory (DFT) to uncover the most favorable decomposition pathways of Ge-rich GST alloys. The knowledge of the DFT formation energy of all GST alloys in the central part of the Ge-Sb-Te ternary phase diagram allowed us to identify the cubic crystalline phases that are more likely to form during the crystallization of a generic GST alloy. This scheme is exemplified by drawing a decomposition map for alloys on the Ge-$Ge_1Sb_2Te_4$ tie-line. A map of decomposition propensity is also constructed, which suggests a possible strategy to minimize phase separation by still keeping a high crystallization temperature.

Keywords: phase change materials; embedded electronic memories; Density Functional Theory; high-throughput calculations

1. Introduction

GeSbTe (GST) phase change alloys have been deeply investigated over the last two decades for a wide range of applications ranging from non-volatile electronic memories (phase change memories, PCMs) [1–4] to neuromorphic computing [5,6], optical disks [7] and several other optical devices [8,9]. These applications rest on two peculiar properties of GST alloys: a reversible and very rapid transition between the amorphous and crystalline phases upon heating and a large contrast in the electronic and optical properties between the two phases, which allows the encoding of the digital information.

In PCMs, the phase transition is induced by Joule heating. During reset, the crystalline phase is rapidly brought above the melting temperature by intense and short current pulses and the amorphous phase is then recovered when the bias is removed due to fast heat dissipation in the surrounding materials. During set, longer and less intense current pulses at a voltage above the Ovonic threshold [10] raise the temperature in between the glass transition temperature T_g and the melting temperature to induce the recrystallization of the amorphous material [1,3,11,12]. Phase change materials are bad glass formers and rapidly crystallize above T_g, which is often identified with the crystallization temperature T_X [13].

The prototypical phase change compound used in PCMs is the $Ge_2Sb_2Te_5$ (GST225) alloy which can be seen as a pseudobinary compound lying on the GeTe-Sb_2Te_3 tie-line. In memory operation, the amorphous phase of GST225 crystallizes into a metastable rocksalt phase and not in the thermodynamically stable trigonal phase due to kinetic hindrances. Note that in the following we will indicate with GSTXYZ the alloy with composition $Ge_xSb_yTe_z$. The amorphous phase of GST alloys is obviously metastable below T_X and it is subject to aging with the increase of the electrical resistance with time [14]. Recrystallization below T_X is also possible on a longer time scale due to the stochasticity of the crystal nucleation process, which leads to data loss of the memory [3]. Data retention thus requires minimizing these unwanted events by suitably tuning the activation energy for the nucleation process at temperatures below T_g.

Although PCMs based on GST alloys are at an advanced stage of development and have reached the global market [15], the exploration of materials in this class is still extensively pursued to enlarge their field of applicability. For instance, GST alloys with T_X higher than that of GST225 are needed for application in embedded memories of interest for the automotive sector [16]. To this end, Ge-rich GST alloys are emerging as promising materials with T_X raising up to 570 K by increasing the Ge content along the Ge-Sb_2Te_3 pseudobinary line [17,18] as opposed to a value T_X in the range 420–440 K for GST225 [19]. Ge-rich alloys along the Ge-$Ge_2Sb_2Te_5$ pseudobinary line [20] or on the Sb-GeTe isoelectronic line (the same average number of p electrons per atom) [21] have also been explored to improve data retention. The high T_X of Ge-rich alloys has been ascribed to the slow-down of the crystallization kinetics due to the mass transport involved in the phase separation into crystalline Ge and less Ge-rich crystalline alloys [22]. In fact, the segregation phenomena require long length diffusion of the atomic species, which would imply a longer incubation time for the formation of supercritical crystalline nuclei. This process was reported in the crystallization of the as-deposited amorphous films and during forming (initialization) of the memory cells [17,18,22–26]. The presence of a reversible phase separation during set/reset is instead less clear. The inhomogeneity of the active material resulting from phase separation has, however, some drawbacks such as a resistance drift in the set (crystalline) state as well, a cell-to-cell variability and possibly reduced endurance [16]. There is therefore room for improvement in the choice of the best composition of the alloy to minimize these detrimental effects. To this end, better knowledge of the decomposition process than is actually available is mandatory. The decomposition process is highly non trivial because, during the operation of the memory, the amorphous phase is supposed to crystallize into metastable cubic phases due to kinetic hindrances, as occurs for GST225, and not necessarily into the thermodynamically stable compounds, which encompass only the unary elements and the pseudobinary compounds along the GeTe-Sb_2Te_3 and Sb-Sb_2Te_3 tie-lines (including the end points) on the Ge-Sb-Te ternary phase diagram.

Atomistic simulations can provide useful information in this respect. In a very recent work, for instance, molecular dynamics simulations based on Density Functional Theory (DFT) were applied to study the amorphous phase of Ge-rich GST along the Ge-GST124 pseudobinary line [27]. The simulations revealed that the amorphous phase is stable with respect to phase separation into amorphous Ge and amorphous GST124 only for Ge content below or equal to 50 atomic %. Phase separation during crystallization was, however, not addressed in Ref. [27].

In this work, we fill this gap by studying the decomposition process during the crystallization of GST alloys by means of high-throughput DFT calculations. In particular, we have constructed the convex hull from the calculations of the DFT total energy of the stable compounds. For all GST cubic alloys in the central part of the ternary phase diagram, we then obtained the distance of their free energy from the convex hull which gives a measure of the metastability of the alloy. These data allowed us to compute the reaction free energy for the decomposition of Ge-rich GST alloys into the crystalline Ge (and eventually Sb and Te), a less Ge-rich alloy, and the GeTe and Sb_2Te_3 binary compounds. The calculations provide a map for the decomposition paths of Ge-rich alloys of any composition and allow us to quantify the propensity to decompose of any alloy and its most probable decomposition products. This approach is exemplified by studying the decomposition pathways of Ge-rich alloys on the Ge-GST124 pseudobinary line. It turns out that several decomposition channels compete, which can lead to the formation of a nanocomposite of crystalline grains with different compositions.

2. Computational Details

We performed DFT calculations by employing the exchange and correlation functional due to Perdew, Burke and Ernzerhof (PBE) [28] and the norm-conserving pseudopotentials by Goedeker, Teter and Hutter [29,30] within the Quickstep method as implemented in the CP2k package [31]. The Kohn–Sham orbitals were expanded in Gaussian-type orbitals of a valence triple-zeta-valence plus polarization basis set while the electronic density was expanded in a basis set of plane waves up to a kinetic energy cutoff of 100 Ry. We also included the van der Waals (vdW) interactions, which have been shown to affect the structural properties of GST compounds [32,33], by including the semiempirical correction due to Grimme [34]. We modeled all GSTXYZ alloys in the cubic rocksalt phase for compositions in the central part of the Ge-Sb-Te ternary phase diagram in which a single element (Ge, Sb or Te) in the alloy does not exceed 60 atomic %. This choice is motivated by the fact that for higher Ge, Sb or Te content we might expect to have either defected tetrahedral geometry, the trigonal layered structures of Sb or more complex structures for very high Te fraction. Since we are interested in the possible switching between the amorphous phase and a cubic crystalline phase, this restriction to the central part of the ternary phase diagram is well justified.

The metastable cubic phase for all compositions was modeled in 216-atom supercells in the rocksalt geometry analogous to the metastable cubic phase of GST225. In cubic GST225, Sb and Ge atoms occupy randomly the cation sublattice with 20% of stoichiometric vacancies, while Te atoms fully occupy the anionic sublattice. Stoichiometric vacancies in GST225 ensure the presence of exactly three p electrons per site on average, which leads to a closed shell system [35].

For GSTXYZ alloys with an atomic fraction of Ge larger than 50%, the cationic sublattice is occupied by Ge only and the anionic sublattice is occupied randomly by Sb, Ge and Te. For Sb-rich alloys, antimony is supposed to behave as an amphoteric element by occupying both the anionic and cationic sublattices. For compositions with more than three p electrons per site, we included vacancies in the cationic sublattice to enforce the condition of three p electrons per site on average. To properly include disorder on the cationic or on both the cationic and anionic sublattices, three different models for each composition were generated by using special quasi-random structures (SQS) [36] with the package of Ref. [37]. The ordered compounds Ge, Te ($P3_121$ space group) [38], Sb ($R\bar{3}m$ space group) [39], GeTe ($R3m$ space group) [40] and Sb_2Te_3 ($R\bar{3}m$ space group) [41] were modeled with 216, 192, 216, 162 and 270 atom supercells. The atomic positions and cell edges of all models were fully optimized with a convergence threshold of 5 mRy/a_0 for forces and a pressure tolerance of 0.01 GPa. The cell of the metastable phases was first optimized by keeping the cubic symmetry of the supercell, and subsequently also the cell angles were allowed to change. Cell optimizations were performed by restricting the Brillouin zone (BZ) integration to the supercell Γ point. The total energy of the models was then computed with a $3 \times 3 \times 3$ k-point mesh in the supercell BZ. In order to estimate the reaction free energies, we also included the configurational entropy in the crystalline phases due to disorder as given by $S = \frac{-k_B}{2} \Sigma_{i,j} x_{i,j} \ln(x_{i,j})$, where the j index runs over the sublattices (i.e., cationic and anionic) and the i index runs over the atomic species which occupy a particular sublattice, k_B is the Boltzmann constant and $x_{i,j}$ is the molar fraction. In the following, we refer to the (reaction) free energy, which includes the total energy at zero temperature and the configurational free energy at 300 K. Calculations on selected possible decomposition pathways have shown that the phononic contribution to the reaction free energy is very small (a few meV/atom) [42] and it has thus been omitted in the data reported hereafter.

3. Results

We first computed the formation free energy of GSTXYZ alloys with respect to the Sb, Ge and Te elements in their standard state. The map of the formation free energy for cubic alloys in the ternary phase diagram is shown in Figure 1. Each point on the map corresponds to a different composition for a total number of 698 alloys uniformly spaced in the central part of the phase diagram. We remind that the free energy is obtained from the total energy at zero temperature plus the configurational free energy at 300 K due to disorder averaged over three SQS models for each composition (see Section 2). A positive formation free energy means that the alloy is unstable with respect to phase separation into the elements Sb, Ge and Te. This is the case for Sb-rich alloys on the left part of the phase diagram. GeSb itself does not exist as a thermodynamically stable compound although a metastable, tetragonally distorted rocksalt phase was reported [43]. The formation free energy is lower (more negative) for alloys along the GeTe-Sb_2Te_3 tie-line and on the Sb-GeTe isoelectronic line. Note that alloys around Sb_2Te_3 are all in the cubic phase as well. A metastable rocksalt phase of Sb_2Te_3 with vacancies on 1/3 of the cationic sites was also recently found experimentally [44].

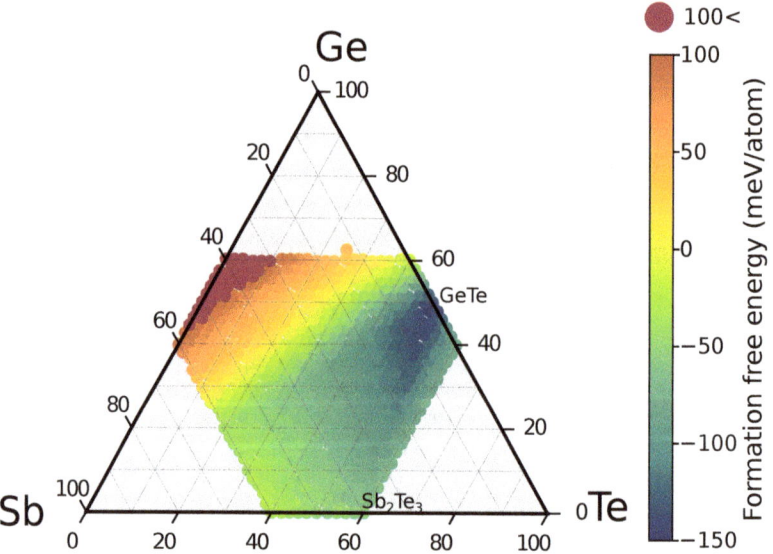

Figure 1. Map of the formation free energy of GST alloys in the metastable cubic phase (at 300 K, see text). The GeTe compound is instead in the stable trigonal phase. The topmost point corresponds to GST512.

The energy of the unary systems and of the trigonal binary compounds GeTe and Sb_2Te_3 is used to construct the convex hull, which, for a ternary system, is the two dimensional surface formed by the ordered compounds that have an energy lower than that of any other structure or any linear combination of structures that provide the proper composition. Compositions at the vertices of the convex hull are thus thermodynamically stable compounds while the tie-lines connecting two compounds are the edges of the convex hull. All other structures have an energy that falls above the set of tie-lines [45–47]. The convex hull represents the Gibbs free energy of the alloy at zero temperature.

Actually, several pseudobinary $(GeTe)_n(Sb_2Te_3)_m$ alloys [48–50] and Sb_nTe compounds [51] are known to be thermodynamically stable and they thus represent vertices of the convex hull. However, the energies of $(GeTe)_n(Sb_2Te_3)_m$ compounds, such as GST147, GST124, GST225, GST326, GST528 and so forth, are very close to the (GeTe)-(Sb_2Te_3) edge of the convex hull. The same is true for the Sb_nTe compounds along the Sb-Sb_2Te_3 tie-line [51].

Therefore, we considered a simplified convex hull including only Ge, Sb, Te and the binary GeTe and Sb_2Te_3 trigonal compounds, which was built by using the qhull code [52]. The geometry of the convex hull with the free energy points of the different cubic alloys is sketched in Figure 2a. A map of the distance of the free energy of the different cubic alloys from the hull is shown instead in Figure 2b. The shorter the distance, the higher the degree of the metastability of the alloy.

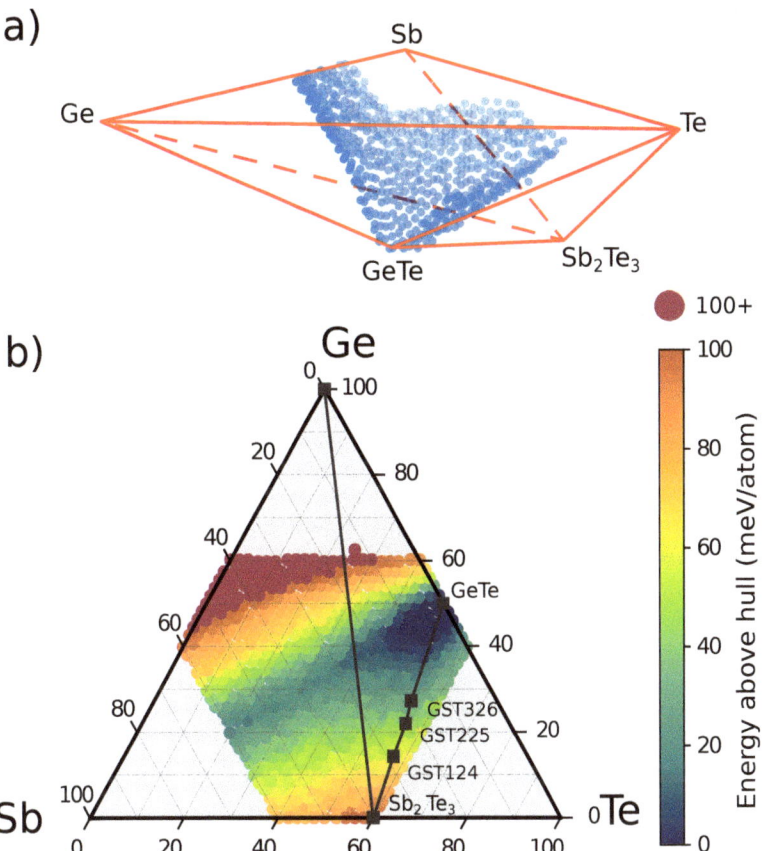

Figure 2. (a) Convex hull in the Ge-Sb-Te phase diagram. The energy of the elements in their standard state is set to zero. The figure shows a simplified convex hull in which the stable pseudobinary GeTe-Sb_2Te_3 compounds and the stable Sb_nTe compounds have been brought on the GeTe-Sb_2Te_3 and Sb-Sb_2Te_3 tie-lines (see text). The formation free energy of the different alloys is also shown by blue points only for compositions with a negative formation free energy (see Figure 1). (b) Map of the distance of the formation free energy (meV/atom) of the cubic alloys from the convex hull. The shorter the distance the higher is the degree of metastability of the alloy. The GeTe-Sb_2Te_3 and Ge-Sb_2Te_3 tie-lines are also shown in panel (b) along with the points corresponding to some stable $(GeTe)_n(Sb_2Te_3)_m$ pseudobinary compounds.

Note that the alloys on the Sb-GeTe isoelectronic line are particularly stable since their energy is very close to that of the convex hull. In fact, the GST212 alloy on the Sb-GeTe line was indicated a few years ago as the starting point to obtain, by Ge enrichment, a "golden composition" for embedded memories [21]. The high metastability (low distance from the convex hull) corresponds to a low propensity to decompose as we will discuss later on. Although cubic alloys along the GeTe-Sb_2Te_3 tie line have a large and negative formation energy (see Figure 1) their distance from the convex hull is larger than that of alloys on the Sb-GeTe line. In fact, the edge of the convex hull along the GeTe-Sb$_2$Te$_3$ tie-line is very close to the energy of the trigonal phases which means that the cubic phases are about 50–60 meV/atom higher in energy than the trigonal ones (see also the very recent DFT work of Ref. [53]).

The map of the distances from the convex hull allows us to study the possible decomposition pathways of any alloy in the phase diagram as follows. We consider the decomposition free energy of the GSTXYZ cubic alloy according to the reaction:

$$GSTXYZ \rightarrow a\,GSTX'Y'Z' + b\,Sb + c\,Ge + d\,Te + e\,Sb_2Te_3 + f\,GeTe. \qquad (1)$$

The GSTX'Y'Z' alloy and the Sb$_2$Te$_3$ compound are in the cubic phase as well, while the other unary systems and the binary GeTe compound are in their ground states (zero temperature). For a given ternary reactant and product the decomposition path is assigned by first maximizing the fraction (a) of the ternary product and then the fraction (e or f) of the binary products. These two constraints uniquely assign the fraction of the unary products. For each GSTXYZ alloy we can then construct a map of the decomposition free energy into a generic GSTX'Y'Z' cubic alloy in the ternary phase diagram. This map highlights which decomposition paths are more likely to be seen during the crystallization of the amorphous phase. Note that the energy of the amorphous phase is always higher than the energy of the cubic phase at the same composition, as we verified with DFT calculations for selected compositions whose amorphous 216-atom models were generated by quenching from the melt following the same protocol we used in previous works [54–57].

As an example of this methodology, we have computed the decomposition map of the three alloys, GST312, GST412 and GST512, on the Ge-GST124 pseudobinary line, whose amorphous phase was investigated by DFT simulations in Ref. [27].

The decomposition maps for the three alloys are compared in Figure 3. A point for the generic GSTX'Y'Z' alloy in the decomposition map of GST312, for instance, gives the value of the decomposition free energy for the formation of GSTX'Y'Z' according to Equation (1). For example, the reaction free energy for the formation of GST334 from GST312 corresponds to the reaction GST312 $\rightarrow$ 1/3 GST334 + 2/3 GeTe + 4/3 Ge. Since the alloys are modeled by a 216-atom supercell, the actual composition is not exactly given by the reaction above but by Ge$_{107}$Sb$_{36}$Te$_{73}$ $\rightarrow$ $\frac{36}{62}$ Ge$_{64}$Sb$_{62}$Te$_{84}$ + $\frac{751}{31}$ GeTe + $\frac{1414}{31}$ Ge, due to the constraints imposed by the finite size of the supercell.

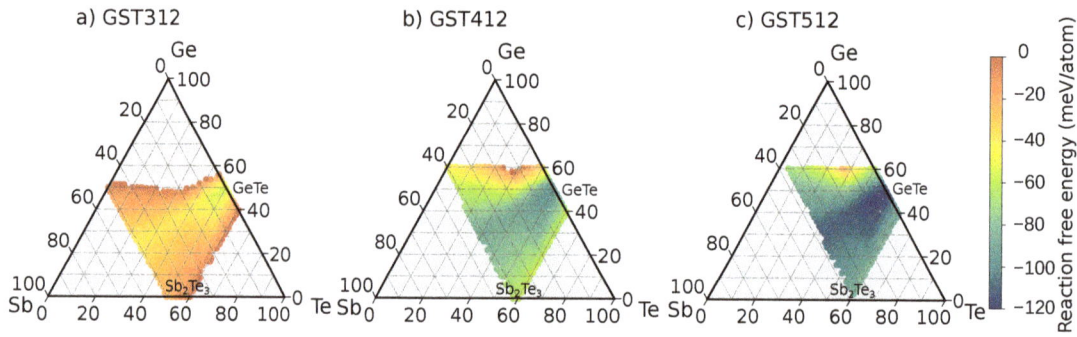

Figure 3. Maps of the decomposition pathways during crystallization of (**a**) GST312, (**b**) GST412, and (**c**) GST512. The color code gives the reaction free energy (meV/atom) to form the cubic alloys on the ternary phase diagram starting from the reactant, i.e., GST312 in panel (**a**) (see text).

A negative reaction free energy indicates an exothermic reaction. The larger and more negative the reaction free energy is, the more favored is the corresponding decomposition pathway.

In the maps shown in Figure 3, we report only exothermic reactions which form GSTX′Y′Z′ in an amount that corresponds at least to 33 atomic % (1/3) of the reactant. Only under these conditions do we consider GSTX′Y′Z′ the main product of the decomposition pathway and show a point on the decomposition map. The composition of the reactant GSTXYZ obviously sets constraints on the possible products, irrespective of the reaction free energy. On the other end, the reaction free energy becomes larger and more negative for a wider part of the phase diagram (the products) by increasing the fraction of Ge (compares GST312 with GST512 in Figure 3). Note that most exothermic reactions lead to the formation of alloys on the Sb-GeTe and GeTe-Sb_2Te_3 lines with a larger weight for alloys close to GeTe.

It is clear that there are several competitive channels for decomposition with very similar reaction free energies for all the three alloys considered in Figure 3 (GST312, GST412, and GST512). The GSTX′Y′Z′ products of these reactions can hardly be discriminated by X-ray diffraction data. The lattice parameter is in fact very similar for the majority of the alloys in the cubic phase as shown in Figure 4, which reports a map of the deviation of the equilibrium lattice parameter from that of GST225 for the different cubic alloys in the ternary phase diagram.

We thus expect the coexistence of different cubic alloys resulting from the crystallization of the amorphous phase of Ge-rich GST alloys during the operation of the memory, and possibly also a cell-to-cell variability due to the presence of several competitive channels for decomposition. The decomposition propensity is, however, different for the three alloys on the Ge-GST124 pseudobinary line investigated here. The alloy that is richer in Ge is in fact more prone to decompose into crystalline Ge and a less Ge-rich alloy, as shown by the presence of a larger blue (more exothermic reactions) region in the decomposition map of GST512. We also computed the decomposition maps of the three alloys by considering both GeTe and Sb_2Te_3 in the trigonal phase. These new maps are shown in Figure A1 in the Appendix A; they are nearly indistinguishable from those of Figure 3 because for Ge-rich alloys there are very few decomposition channels leading to crystalline Sb_2Te_3.

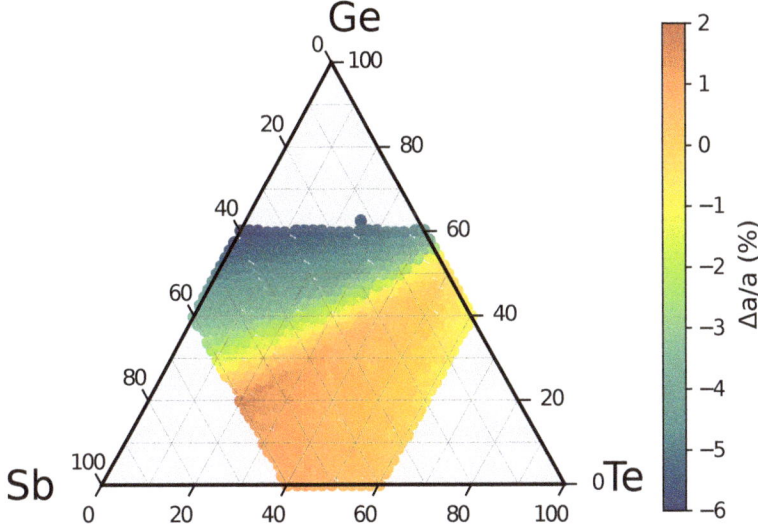

Figure 4. Map of the lattice parameter of the cubic phase of GST alloys in the ternary phase diagram. The lattice parameter is shown as the relative difference $(a - a_{225})/a_{225}$ in percentage (%) with respect to the lattice parameter of GST225 (a_{225} = 6.0293 Å) [49].

We then attempted to quantify the decomposition propensity by counting the exothermic decomposition channels weighted by their reaction free energy. In this calculation, we now include all the possible products in the central region of the ternary phase diagram (all the 698 equally spaced alloys) even if they appear in a small atomic fraction in the decomposition product. This is done because a reaction with a low fraction of the GSTX′Y′Z′ products can still be strongly exothermic due to the formation of a large fraction of GeTe. This procedure is repeated now for all possible starting GSTXYZ alloys and we then obtain the map of the propensity to decompose for alloys in the central region of the phase diagram shown in Figure 5. The decomposition propensity is low again along the GeTe-Sb_2Te_3 tie-line but also along the Sb-GeTe isoelectronic line. This map suggests a possible tuning of the alloy composition to minimize the decomposition propensity by still keeping a higher T_X, as we discuss below.

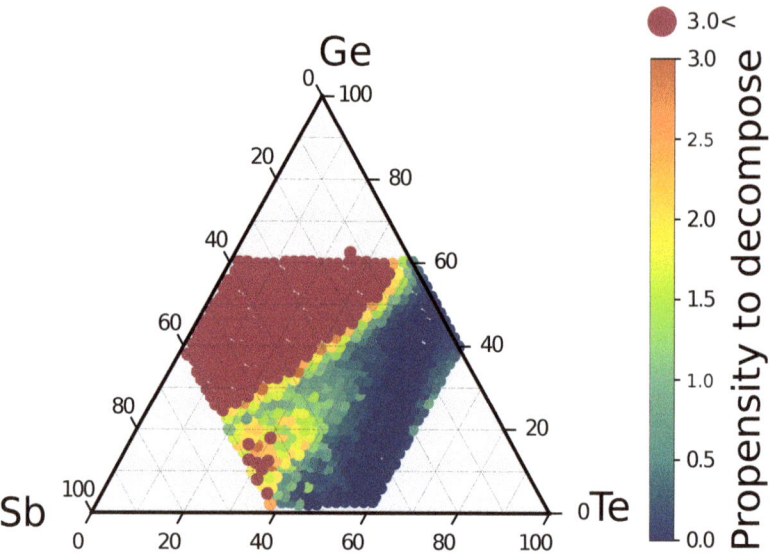

Figure 5. Map of the decomposition propensity of the cubic phase of GST alloys on the ternary phase diagram (see text). Low values (blue region) indicate a low propensity to decompose. The decomposition propensity is computed by counting the exothermic decomposition channels weighted by the modulus of their reaction free energy (in eV/atom).

In principle, it would be desirable to select an alloy with a sufficiently high T_X that still switches between the crystalline and amorphous phases in a homogeneous manner with no phase separation. This would allow the minimization of the cell-to-cell variability and the reduction of the drift of the set state that could arise from the presence of a partially crystallized system or of nanocrystallites with different compositions that may undergo coarsening. The phase separation itself is believed to be the origin of the raise of T_X because of the mass transport on a long length scale involved in phase segregation [22]. However, previous analysis of DFT models of the amorphous phase as a function of composition revealed that moving away from the GeTe-Sb_2Te_3 tie-line by increasing either the Ge or Sb content along the Ge-GST124 [27], Sb-GeTe [56] and Ge-Sb_2Te_3 lines [42], leads to an amorphous network which is more and more dissimilar from the cubic crystal. In the rocksalt phase, all atoms are in an octahedral bonding geometry with no homopolar bonds. On the contrary, in the amorphous phase, a fraction of Ge atoms are in a tetrahedral bonding geometry favored by the presence of Ge–Ge bonds [54]. The fraction of Ge–Ge bonds and of tetrahedra increases by increasing the Ge content but also the Sb fraction because of the lack of a sufficient amount of Te to form Ge–Te bonds [27,56]. Another indicator of the bonding topology in the amorphous phase is given by the distribution of rings, that is, the shortest closed loops among atoms connected by bonds. In GST225 and GeTe, the ring distribution is dominated by four-membered ABAB rings [58,59] (A = Ge/Sb, B = Te), which are the building blocks of the cubic rocksalt crystal. The four-membered rings are present in the amorphous phase but also in the supercooled liquid phase above T_g where crystallization takes place [59]. Molecular dynamics simulations of the crystallization process [59–61] have shown that the reorientation of the four-membered rings is the key process in the formation of cubic crystalline nuclei. By increasing the Ge and Sb content at the expense of Te, the fraction of four-membered rings reduces while the number of five- and six-membered rings increases [27,42,56]. Homopolar bonds, tetrahedra and longer rings in place of four-membered ones are all structural features hindering the crystallization since they are not present in the crystal [3,27,42,56]. Therefore, we might conceive to reduce the crystal nucleation rate and then improve data retention by increasing the Ge or Sb fraction to enhance the dissimilarity between the crystal and amorphous phases by moving away from the GeTe-Sb$_2$Te$_3$ pseudobinary line. The propensity map in Figure 5 tells us that if we increase the Ge or Sb content by keeping close to the Sb-GeTe line we can also minimize the propensity to decompose during crystallization. It remains to be seen if this strategy is suitable to raise T_X sufficiently for applications in embedded memories.

As a final result, we report in Figure 6 a map of the average atomic fraction of segregated Ge along the decomposition pathways of a generic GSTXYZ alloy in the ternary phase diagram. This average fraction is computed by summing the atomic fraction of Ge (in the crystalline form) formed in each exothermic decomposition path weighted by its reaction free energy Δ as $Ge_{av}(GSTXYZ) = \frac{\sum_i c_i |\Delta_i|}{X \sum_i |\Delta_i|}$, where the sum runs over the decomposition channels and c_i is the fraction of Ge formed in reaction i (see Equation (1)). As expected, there is a rough correspondence between the average fraction of segregated Ge (Figure 6) and the propensity to decompose (Figure 5), but the map of the decomposition propensity shows some finer details (e.g., a lower decomposition propensity along the Sb–GeTe line) ,which are not present in the map of Figure 6.

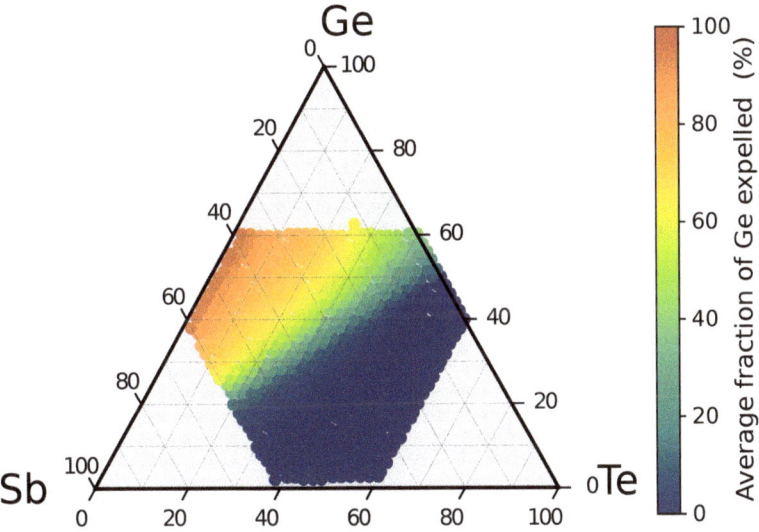

Figure 6. Map of the average fraction of crystalline Ge in atomic % segregated in the exothermic decomposition reactions of GST alloys in the central part of the ternary phase diagram (see text for definition).

4. Discussion and Conclusions

The analysis of the decomposition pathways suggests that the crystallization of Ge-rich alloys could lead to the formation of less Ge-rich cubic crystals with different compositions. The coexistence of different competitive decomposition channels might affect the cell-to-cell variability and the drift in the set state of the memory. However, we must remark that the decomposition maps in Figure 3 correspond to the final result of the decomposition reaction with all products in the crystalline state. This process is, however, likely to occur in different steps. For instance, in the real samples, Ge could migrate to enrich the amorphous region surrounding the crystallizing region depleted in Ge. The crystallization of Ge inside the further Ge enriched amorphous region will occur at a later stage or at higher temperature [26].

Therefore, although the decomposition maps for the individual alloy (Figure 3) and the map of the decomposition propensity (Figure 5) suggest possible strategies to mimimize phase segregation, the result of the actual crystallization process is expected to strongly depend on the details of the thermal history of the sample. In fact, it was already noted that the kinetics of crystallization are different for the as-deposited and melt quenched amorphous phase annealed above T_g, and it is also different from phase separation occurring by directly quenching the liquid below melting [24]. Furthermore, in a nanoscaled device, the switching between the amorphous and the crystalline phases occurs only in a small dome above the heater in the mushroom architecture. The nature of the surrounding material might also affect the expulsion of Ge from the crystallizing region into the outer part of the transforming dome. The forming operation is indeed known to affect the switching process because it controls the composition of both the transforming dome and the surrounding region [18]. The possible formation of a nanocomposite made of crystalline grains of different compositions is also expected to be strongly affected by nanoconfinement; for instance, in a multilayer geometry such as that explored with $Sb_2Te_3/TiTe_2$ heterostructures for neuromorphic applications in Ref. [62]. We thus simply mention that nanostructuring Ge-rich GST alloys in multilayers might represent another possible strategy for mitigating the phase separation.

In summary, by means of high-throughput DFT calculations we have provided a comprehensive analysis of the possible decomposition pathways that GST amorphous alloys could undergo during crystallization. This analysis focuses in particular on Ge-rich GST alloys of interest for applications in embedded memories for the automotive sector that require high crystallization temperatures.

The calculation of the reaction free energy for the phase separation of crystalline Ge-rich GST into crystalline Ge and less Ge-rich cubic alloys allowed us to identify the thermodynamically more viable decomposition paths. Other kinetic effects related to the process of segregation of Ge might, however, affect the actual products of the phase separation as discussed above. Although these effects cannot be addressed by DFT methods, the recent developments of machine learning interatomic potentials for large scale simulations [63–65] (several thousands of atoms for tens of ns), eventually supplemented by accelerating techniques such as the metadynamics method [66,67], pave the way for the direct molecular dynamics simulations of the crystallization of Ge-rich GST alloys to enlighten the intermediate steps of the whole decomposition process.

Author Contributions: Both authors equally contributed to the work. Both authors have read and agreed to the published version of the manuscript.

Funding: This work was partly funded by the European Union Horizon 2020 research and innovation program under Grant Agreement No. 824957 (BeforeHand: Boosting Performance of Phase Change Devices by Hetero- and Nanostructure Material Design). We thankfully acknowledge the computational resources provided by the ISCRA program at Cineca (Casalecchio di Reno, Italy).

Data Availability Statement: The data that support the findings of this study are available from the corresponding author upon reasonable request.

Conflicts of Interest: The authors declare no conflict of interest.

Abbreviations

The following abbreviations are used in this manuscript:

GST	GeSbTe
PCMs	phase change memories
GSTXYZ	$Ge_X Sb_Y Ye_Z$
DFT	Density Functional Theory
PBE	Perdew, Burke and Ernzerhof
vdW	van der Waals
BZ	Brillouin zone
SQS	Special quasi-random structure

Appendix A

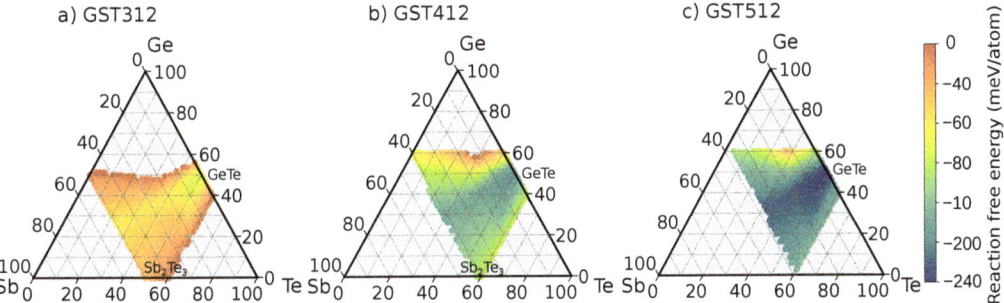

Figure A1. Maps of the decomposition pathways during crystallization of (**a**) GST312, (**b**) GST412, and (**c**) GST512. The color code gives the reaction free energy (meV/atom) to form the cubic alloys on the ternary phase diagram starting from the reactant, i.e., GST312 in panel (**a**) (see text). In these maps the Sb_2Te_3 product is considered in the trigonal phase as opposed to the maps in Figure 3 where the cubic phase was used for Sb_2Te_3 as well.

References

1. Wuttig, M.; Yamada, N. Phase-change materials for rewriteable data storage. *Nat. Mater.* **2007**, *6*, 824–832. [CrossRef]
2. Noé, P.; Vallée, C.; Hippert, F.; Fillot, F.; Raty, J.Y. Phase-change materials for non-volatile memory devices: From technological challenges to materials science issues. *Semicond. Sci. Technol.* **2018**, *33*, 013002. [CrossRef]
3. Zhang, W.; Mazzarello, R.; Wuttig, M.; Ma, E. Designing crystallization in phase-change materials for universal memory and neuro-inspired computing. *Nat. Rev. Mater.* **2019**, *4*, 150–168. [CrossRef]
4. Pirovano, A.; Lacaita, A.; Benvenuti, A.; Pellizzer, F.; Bez, R. Electronic switching in phase-change memories. *IEEE Trans. Electron Devices* **2004**, *51*, 452–459. [CrossRef]
5. Boybat, I.; Le Gallo, M.; Nandakumar, S.R.; Moraitis, T.; Parnell, T.; Tuma, T.; Rajendran, B.; Leblebici, Y.; Sebastian, A.; Eleftheriou, E. Neuromorphic computing with multi-memristive synapses. *Nat. Commun.* **2018**, *9*, 1–12. [CrossRef]
6. Sebastian, A.; Le Gallo, M.; Khaddam-Aljameh, R.; Eleftheriou, E. Memory devices and applications for in-memory computing. *Nat. Nanotechnol.* **2020**, *15*, 529–544. [CrossRef]
7. Satoh, I.; Yamada, N. DVD-RAM for all audio/video, PC, and network applications. In *Fifth International Symposium on Optical Storage (ISOS 2000)*; Gan, F., Hou, L., Eds.; International Society for Optics and Photonics, SPIE: Bellingham, WA, USA, 2001; Volume 4085, pp. 283–290. [CrossRef]
8. Feldmann, J.; Youngblood, N.; Karpov, M.; Gehring, H.; Li, X.; Stappers, M.; Le Gallo, M.; Fu, X.; Lukashchuk, A.; Raja, A.S.; et al. Parallel convolutional processing using an integrated photonic tensor core. *Nature* **2021**, *589*, 52–58. [CrossRef] [PubMed]
9. Wuttig, M.; Bhaskaran, H.; Taubner, T. Phase-change materials for non-volatile photonic applications. *Nat. Photon.* **2017**, *11*, 465–476. [CrossRef]
10. Adler, D.; Shur, M.; Silver, M.; Ovshinsky, S. Threshold switching in chalcogenide-glass thin films. *J. Appl. Phys.* **1980**, *51*, 3289–3309. [CrossRef]
11. Raoux, S.; Wełnic, W.; Ielmini, D. Phase Change Materials and Their Application to Nonvolatile Memories. *Chem. Rev.* **2010**, *110*, 240–267. [CrossRef] [PubMed]
12. Gallo, M.L.; Sebastian, A. An overview of phase-change memory device physics. *J. Phys. D Appl. Phys.* **2020**, *53*, 213002. [CrossRef]

13. Pries, J.; Wei, S.; Wuttig, M.; Lucas, P. Switching between Crystallization from the Glassy and the Undercooled Liquid Phase in Phase Change Material Ge$_2$Sb$_2$Te$_5$. *Adv. Mater.* **2019**, *31*, 1900784. [CrossRef] [PubMed]
14. Zhang, W.; Ma, E. Unveiling the structural origin to control resistance drift in phase-change memory materials. *Mater. Today* **2020**, *41*, 156–176. [CrossRef]
15. Choe, J. Intel 3D XPoint Memory Die Removed from Intel Optane™ PCM (Phase Change Memory). 2017. Available online: https://www.techinsights.com/blog/intel-3d-xpoint-memory-die-removed-intel-optanetm-pcm-phase-change-memory (accessed on 7 August 2021).
16. Cappelletti, P.; Annunziata, R.; Arnaud, F.; Disegni, F.; Maurelli, A.; Zuliani, P. Phase change memory for automotive grade embedded NVM applications. *J. Phys. D Appl. Phys.* **2020**, *53*, 193002. [CrossRef]
17. Zuliani, P.; Palumbo, E.; Borghi, M.; Dalla Libera, G.; Annunziata, R. Engineering of chalcogenide materials for embedded applications of Phase Change Memory. *Solid State Electron.* **2015**, *111*, 27–31. [CrossRef]
18. Palumbo, E.; Zuliani, P.; Borghi, M.; Annunziata, R. Forming operation in Ge-rich Ge$_x$Sb$_y$Te$_z$ phase change memories. *Solid State Electron.* **2017**, *133*, 38–44. [CrossRef]
19. Noé, P.; Sabbione, C.; Bernier, N.; Castellani, N.; Fillot, F.; Hippert, F. Impact of interfaces on scenario of crystallization of phase change materials. *Acta Mater.* **2016**, *110*, 142–148. [CrossRef]
20. Navarro, G.; Coué, M.; Kiouseloglou, A.; Noé, P.; Fillot, F.; Delaye, V.; Persico, A.; Roule, A.; Bernard, M.; Sabbione, C.; et al. Trade-off between SET and data retention performance thanks to innovative materials for phase-change memory. In Proceedings of the 2013 IEEE International Electron Devices Meeting, Washington, DC, USA, 9–11 December 2013; pp. 21.5.1–21.5.4. [CrossRef]
21. Cheng, H.Y.; Hsu, T.H.; Raoux, S.; Wu, J.; Du, P.Y.; Breitwisch, M.; Zhu, Y.; Lai, E.K.; Joseph, E.; Mittal, S.; et al. A high performance phase change memory with fast switching speed and high temperature retention by engineering the Ge$_x$Sb$_y$Te$_z$ phase change material. In Proceedings of the 2011 International Electron Devices Meeting, Washington, DC, USA, 5–7 December 2011; pp. 3.4.1–3.4.4. [CrossRef]
22. Agati, M.; Vallet, M.; Joulié, S.; Benoit, D.; Claverie, A. Chemical phase segregation during the crystallization of Ge-rich GeSbTe alloys. *J. Mater. Chem. C* **2019**, *7*, 8720–8729. [CrossRef]
23. Privitera, S.M.S.; Sousa, V.; Bongiorno, C.; Navarro, G.; Sabbione, C.; Carria, E.; Rimini, E. Atomic diffusion in laser irradiated Ge rich GeSbTe thin films for phase change memory applications. *J. Phys. D Appl. Phys.* **2018**, *51*, 145103. [CrossRef]
24. Privitera, S.M.; López García, I.; Bongiorno, C.; Sousa, V.; Cyrille, M.C.; Navarro, G.; Sabbione, C.; Carria, E.; Rimini, E. Crystallization properties of melt-quenched Ge-rich GeSbTe thin films for phase change memory applications. *J. Appl. Phys.* **2020**, *128*, 155105. [CrossRef]
25. Henry, L.; Bernier, N.; Jacob, M.; Navarro, G.; Clément, L.; Rouvière, J.L.; Robin, E. Studying phase change memory devices by coupling scanning precession electron diffraction and energy dispersive X-ray analysis. *Acta Mater.* **2020**, *201*, 72–78. [CrossRef]
26. Luong, M.; Agati, M.; Ramond, N.; Grisolia, J.; Le Friec, Y.; Benoit, D.; Claverie, A. On Some Unique Specificities of Ge-Rich GeSbTe Phase-Change Material Alloys for Nonvolatile Embedded-Memory Applications. *Phys. Status Solidi RRL* **2021**, *15*, 2000471. [CrossRef]
27. Sun, L.; Zhou, Y.X.; Wang, X.D.; Chen, Y.H.; Deringer, V.L.; Mazzarello, R.; Zhang, W. Ab initio molecular dynamics and materials design for embedded phase-change memory. *npj Comput. Mater.* **2021**, *7*, 29. [CrossRef]
28. Perdew, J.P.; Burke, K.; Ernzerhof, M. Generalized gradient approximation made simple. *Phys. Rev. Lett.* **1996**, *77*, 3865. [CrossRef] [PubMed]
29. Goedecker, S.; Teter, M. Separable dual-space Gaussian pseudopotentials. *Phys. Rev. B Condens. Matter* **1996**, *54*, 1703. [CrossRef] [PubMed]
30. Krack, M. Pseudopotentials for H to Kr optimized for gradient-corrected exchange-correlation functionals. *Theor. Chem. Acc.* **2005**, *114*, 145–152. [CrossRef]
31. Vandevondele, J.; Krack, M.; Mohamed, F.; Parrinello, M.; Chassaing, T.; Hutter, J. Quickstep: Fast and accurate density functional calculations using a mixed Gaussian and plane waves approach. *Comput. Phys. Commun.* **2005**, *167*, 103–128. [CrossRef]
32. Weber, H.; Schumacher, M.; Jóvári, P.; Tsuchiya, Y.; Skrotzki, W.; Mazzarello, R.; Kaban, I. Experimental and ab initio molecular dynamics study of the structure and physical properties of liquid GeTe. *Phys. Rev. B* **2017**, *96*, 054204. [CrossRef]
33. Schumacher, M.; Weber, H.; Jóvári, P.; Tsuchiya, Y.; Youngs, T.G.; Kaban, I.; Mazzarello, R. Structural, electronic and kinetic properties of the phase-change material Ge$_2$Sb$_2$Te$_5$ in the liquid state. *Sci. Rep.* **2016**, *6*, 27434. [CrossRef] [PubMed]
34. Grimme, S.; Antony, J.; Ehrlich, S.; Krieg, H. A consistent and accurate ab initio parametrization of density functional dispersion correction (DFT-D) for the 94 elements H-Pu. *J. Chem. Phys.* **2010**, *132*, 154104. [CrossRef]
35. Luo, M.; Wuttig, M. The Dependence of Crystal Structure of Te-Based Phase-Change Materials on the Number of Valence Electrons. *Adv. Mater.* **2004**, *16*, 439–443. [CrossRef]
36. Zunger, A.; Wei, S.H.; Ferreira, L.G.; Bernard, J.E. Special quasirandom structures. *Phys. Rev. Lett.* **1990**, *65*, 353. [CrossRef]
37. Van De Walle, A.; Tiwary, P.; De Jong, M.; Olmsted, D.L.; Asta, M.; Dick, A.; Shin, D.; Wang, Y.; Chen, L.Q.; Liu, Z.K. Efficient stochastic generation of special quasirandom structures. *Calphad* **2013**, *42*, 13–18. [CrossRef]
38. Adenis, C.; Langer, V.; Lindqvist, O. Reinvestigation of the structure of tellurium. *Acta Cryst. C* **1989**, *45*, 941–942. [CrossRef]
39. Barrett, C.S.; Cucka, P.; Haefner, K. The crystal structure of antimony at 4.2, 78 and 298 K. *Acta Cryst.* **1963**, *16*, 451–453. [CrossRef]
40. Goldak, J.; Barrett, C.S.; Innes, D.; Youdelis, W. Structure of α-GeTe. *J. Chem. Phys* **1966**, *44*, 3323–3325. [CrossRef]

41. Anderson, T.L.; Krause, H.B. Refinement of the Sb_2Te_3 and Sb_2Te_2Se structures and their relationship to nonstoichiometric $Sb_2Te_{3-y}Se_y$ compounds. *Acta Cryst. B* **1974**, *30*, 1307–1310. [CrossRef]
42. Abou El Kheir, O.; Dragoni, D.; Bernasconi, M. Density functional simulations of decomposition pathways of Ge-rich GeSbTe alloys for phase change memories. *Phys. Rev. Mat.* **2021**, in press.
43. Giessen, B.; Borromee-Gautier, C. Structure and alloy chemistry of metastable GeSb. *J. Solid State Chem.* **1972**, *4*, 447–452. [CrossRef]
44. Li, K.; Peng, L.; Zhu, L.; Zhou, J.; Sun, Z. Vacancy-mediated electronic localization and phase transition in cubic Sb_2Te_3. *Mater. Sci. Semicond. Process.* **2021**, *135*, 106052. [CrossRef]
45. Hildebrandt, G.; Glasser, D. Predicting phase and chemical equilibrium using the convex hull of the Gibbs free energy. *Chem. Eng. J.* **1994**, *54*, 187–197. [CrossRef]
46. Curtarolo, S.; Morgan, D.; Ceder, G. Accuracy of ab initio methods in predicting the crystal structures of metals: A review of 80 binary alloys. *Comput. Coupling Phase Diagrams Thermochem.* **2005**, *29*, 163–211. [CrossRef]
47. Nyshadham, C.; Oses, C.; Hansen, J.E.; Takeuchi, I.; Curtarolo, S.; Hart, G.L.W. A computational high-throughput search for new ternary superalloys. *Acta Mater.* **2017**, *122*, 483–487. [CrossRef]
48. Matsunaga, T.; Kojima, R.; Yamada, N.; Kifune, K.; Kubota, Y.; Takata, M. Structural Features of $Ge_1Sb_4Te_7$, an Intermetallic Compound in the $GeTe-Sb_2Te_3$ Homologous Series. *Chem. Mater.* **2008**, *20*, 5750–5755. [CrossRef]
49. Matsunaga, T.; Yamada, N.; Kubota, Y. Structures of stable and metastable $Ge_2Sb_2Te_5$, an intermetallic compound in $GeTe-Sb_2Te_3$ pseudobinary systems. *Acta Cryst. B* **2004**, *60*, 685–691. [CrossRef] [PubMed]
50. Matsunaga, T.; Morita, H.; Kojima, R.; Yamada, N.; Kifune, K.; Kubota, Y.; Tabata, Y.; Kim, J.J.; Kobata, M.; Ikenaga, E.; et al. Structural characteristics of GeTe-rich $GeTe-Sb_2Te_3$ pseudobinary metastable crystals. *J. Appl. Phys.* **2008**, *103*, 093511. [CrossRef]
51. Govaerts, K.; Sluiter, M.H.; Partoens, B.; Lamoen, D. Stability of Sb-Te layered structures: First-principles study. *Phys. Rev. B Condens. Matter* **2012**, *85*, 1–8. [CrossRef]
52. Barber, C.B.; Dobkin, D.P.; Huhdanpaa, H. The Quickhull algorithm for convex hulls. *ACM Trans. Math. Softw.* **1996**, *22*, 469–483. [CrossRef]
53. Evang, V.; Mazzarello, R. Point defects in disordered and stable GeSbTe phase-change materials. *Mater. Sci. Semicond. Process.* **2021**, *133*, 105948. [CrossRef]
54. Caravati, S.; Bernasconi, M.; Kühne, T.D.; Krack, M.; Parrinello, M. Coexistence of tetrahedral- and octahedral-like sites in amorphous phase change materials. *Appl. Phys. Lett.* **2007**, *91*, 171906. [CrossRef]
55. Caravati, S.; Bernasconi, M.; Kühne, T.D.; Krack, M.; Parrinello, M. First-principles study of crystalline and amorphous $Ge_2Sb_2Te_5$ and the effects of stoichiometric defects. *J. Condens. Matter Phys.* **2009**, *21*, 255501. [CrossRef]
56. Gabardi, S.; Caravati, S.; Bernasconi, M.; Parrinello, M. Density functional simulations of Sb-rich GeSbTe phase change alloys. *J. Condens. Matter Phys.* **2012**, *24*, 385803. [CrossRef]
57. Sosso, G.C.; Caravati, S.; Mazzarello, R.; Bernasconi, M. Raman spectra of cubic and amorphous $Ge_2Sb_2Te_5$ from first principles. *Phys. Rev. B* **2011**, *83*, 134201. [CrossRef]
58. Akola, J.; Jones, R.O. Structural phase transitions on the nanoscale: The crucial pattern in the phase-change materials $Ge_2Sb_2Te_5$ and GeTe. *Phys. Rev. B* **2007**, *76*, 235201. [CrossRef]
59. Hegedüs, J.; Elliott, S.R. Microscopic origin of the fast crystallization ability of Ge–Sb–Te phase-change memory materials. *Nat. Mater.* **2008**, *7*, 399–405. [CrossRef]
60. Kalikka, J.; Akola, J.; Jones, R.O. Crystallization processes in the phase change material $Ge_2Sb_2Te_5$: Unbiased density functional/molecular dynamics simulations. *Phys. Rev. B* **2016**, *94*, 134105. [CrossRef]
61. Sosso, G.C.; Miceli, G.; Caravati, S.; Giberti, F.; Behler, J.; Bernasconi, M. Fast Crystallization of the Phase Change Compound GeTe by Large-Scale Molecular Dynamics Simulations. *J. Phys. Chem. Lett.* **2013**, *4*, 4241–4246. [CrossRef] [PubMed]
62. Ding, K.; Wang, J.; Zhou, Y.; Tian, H.; Lu, L.; Mazzarello, R.; Jia, C.; Zhang, W.; Rao, F.; Ma, E. Phase-change heterostructure enables ultralow noise and drift for memory operation. *Science* **2019**, *366*, 210–215. [CrossRef] [PubMed]
63. Behler, J. First Principles Neural Network Potentials for Reactive Simulations of Large Molecular and Condensed Systems. *Angew. Chem. Int. Ed.* **2017**, *56*, 12828–12840. [CrossRef] [PubMed]
64. Friederich, P.; Häse, F.; Proppe, J.; Aspuru-Guzik, A. Machine-learned potentials for next-generation matter simulations. *Nat. Mater.* **2021**, *20*, 750–761. [CrossRef]
65. Deringer, V.L.; Caro, M.A.; Csányi, G. Machine Learning Interatomic Potentials as Emerging Tools for Materials Science. *Adv. Mater.* **2019**, *31*, 1902765. [CrossRef] [PubMed]
66. Ronneberger, I.; Zhang, W.; Eshet, H.; Mazzarello, R. Crystallization Properties of the $Ge_2Sb_2Te_5$ Phase-Change Compound from Advanced Simulations. *Adv. Funct. Mater.* **2015**, *25*, 6407–6413. [CrossRef]
67. Laio, A.; Parrinello, M. Escaping free-energy minima. *Proc. Natl. Acad. Sci. USA* **2002**, *99*, 12562–12566. [CrossRef] [PubMed]

Article

Structural Assessment of Interfaces in Projected Phase-Change Memory

Valeria Bragaglia *, Vara Prasad Jonnalagadda, Marilyne Sousa, Syed Ghazi Sarwat, Benedikt Kersting and Abu Sebastian *

IBM Research Europe—Zurich Research Laboratory, CH-8803 Rüschlikon, Switzerland; vjo@zurich.ibm.com (V.P.J.); sou@zurich.ibm.com (M.S.); ghs@zurich.ibm.com (S.G.S.); bke@zurich.ibm.com (B.K.)
* Correspondence: vbr@zurich.ibm.com (V.B.); ase@zurich.ibm.com (A.S.)

Abstract: Non-volatile memories based on phase-change materials have gained ground for applications in analog in-memory computing. Nonetheless, non-idealities inherent to the material result in device resistance variations that impair the achievable numerical precision. Projected-type phase-change memory devices reduce these non-idealities. In a projected phase-change memory, the phase-change storage mechanism is decoupled from the information retrieval process by using projection of the phase-change material's phase configuration onto a projection liner. It has been suggested that the interface resistance between the phase-change material and the projection liner is an important parameter that dictates the efficacy of the projection. In this work, we establish a metrology framework to assess and understand the relevant structural properties of the interfaces in thin films contained in projected memory devices. Using X-ray reflectivity, X-ray diffraction and transmission electron microscopy, we investigate the quality of the interfaces and the layers' properties. Using demonstrator examples of Sb and Sb_2Te_3 phase-change materials, new deposition routes as well as stack designs are proposed to enhance the phase-change material to a projection-liner interface and the robustness of material stacks in the devices.

Keywords: in-memory computing; projected phase-change memory; interface engineering; confined phase-change material; sputtering deposition; X-ray reflectivity; STEM

1. Introduction

Chalcogenide phase-change materials (PC-materials) are well-known for their crystalline-to-amorphous transitions that are both fast and reversible, offering a large electrical and optical contrast between the two phases [1,2]. This class of materials has greatly contributed to the development of highly dense data storage in optical media and to solid-state non-volatile memory [3].

In recent years, phase-change memories (PCMs) have gained ground as non-volatile memories in the field of analog in-memory computing and neuromorphic systems [4–7]. It has been shown that PCM devices can potentially store a continuum of resistance values. When organized in crossbar arrays, PCMs can be used to perform vector-matrix multiplications, the core calculation of artificial intelligence (AI)'s inference and training tasks, by exploiting Ohm's and Kirchoff's laws.

However, initially developed for binary data storage [8,9] rather than for computation, the technology endures significant non-idealities due to the intrinsic material physics of PC-materials. Resistance drift and electrical read noise induce temporal variations in the device resistance [10,11], reducing the numerical precision that is achievable with this technology.

To minimize these non-idealities, various solutions based on both materials and device engineering have been proposed [12–14]. Among the newly proposed cell architectures, the concept of projected PCM has gained much interest [13,15,16]. In this concept, the physical mechanism of resistance storage is decoupled from the noisy information retrieval process that is affected by the electrical properties of the PC-material's amorphous phase.

The decoupling is realized by using an electrically conducting material, called the projection liner (PL), placed in parallel to the PC-material, as illustrated in Figure 1a. The resistance of the PL is judiciously chosen such that it only has a marginal influence on the write operation. In effect, the device read-out characteristics become dictated by the properties of the PL. Projected phase-change memory devices have been shown to significantly reduce resistance drift and 1/f noise by at least one order of magnitude. Such improvements enable in-memory arithmetic operations with high precision [17].

Two recent studies focused on the understanding of the "projection", investigating the device characteristics for line-cell architectures [18] as well as for vertical devices, the so called "mushroom-type phase-change memory devices" [16]. Based on these two studies, a comprehensive device model was developed to capture the behavior of the memory device for any arbitrary device state. For line-cell architectures, it was found that the interface resistance between the PC-material and PL is a crucial parameter, determining how effectively the projection works. It hampers the current flow into the PL and, thus, determines the fraction of read current that bypasses the amorphous PC-material volume [18]. As for a lateral-type projected PCM, the two extreme scenarios for $R_{interface} = 0\ \Omega$ and $R_{interface} = \infty\ \Omega$ are illustrated in Figure 1a,b, respectively. For a "zero-interface resistance", the current mostly bypasses the amorphous PC-material region, leading to reduced non-idealities. If the interface resistance is infinite, the projection current bypasses the entire phase-change layer, eventually leading to a binary cell behavior. Moreover, it was shown that the interface resistance also affects both the drift characteristics and the state dependence of the device resistance. An experimental validation of the findings was also provided for projected PCM with Sb or $Ge_2Sb_2Te_5$ material on top of a metal nitride (MeN) PL [16,18].

Here, we extend the understanding of previous studies based on a new perspective: we establish a metrology to assess relevant structural properties in projected PCM. X-ray reflectivity (XRR), X-ray diffraction (XRD), scanning transmission electron microscopy (STEM) and energy-dispersive X-ray spectroscopy (EDX) are orchestrated to determine the quality of the layer stacks with a focus on the interface and layer attributes. Spurious layers are identified and discussed in the framework of the interfacial resistance and the non-ideal device performances of previous studies [17,19]. The investigation applies to thin Sb as well as Sb_2Te_3 material systems for thicknesses from 10 nm down to the ultra-scaled case of 3 nm.

Several deposition approaches as well as stack variations are proposed to enhance not only the PC-material-to-PL interface but also the uniformity of the PC-material and the robustness of material stacks against thermal stress and a resulting potential mechanical strain upon fabrication. Eventually, we establish the groundwork for the conscious designing of the next generation of projected phase-change memory.

2. Materials and Methods

The various stacks were deposited using an FHR.Star.75.Co sputter tool. This is a multi-target sputtering system with five sputter sources, one of which was used for the inverse sputter etch process. In addition, the tool was fitted with a heating substrate stage that can reach a maximum temperature (T) of 600 °C. It can deposit multiple layers of metallic, phase-change and dielectric thin films in situ at elevated temperatures.

For this study, each layer stack was deposited in situ. This permitted a full preparation without breaking the vacuum and, therefore, prevented the oxidation and contamination of the layer stacks. The deposition parameters for the various layers are summarized in Table 1.

Table 1. Summary of the deposition parameters used for the various samples' layers.

Layer	Deposition Parameters
MeN	Reactive sputtering
Sb	DC sputtering: 50 W, 3 µbar, 110 sccm Ar
Sb_2Te_3	DC sputtering: 60 W, 6 µbar, 110 sccm Ar
SiO_2 capping	RF sputtering: 250 W, 6 µbar, 110 sccm Ar

To analyze the structural properties of the stacks, such as phase, thickness, density and interface quality between layers, XRD and XRR measurements were performed in a Bruker D8 discover diffractometer equipped with a rotating anode generator.

The fitting analysis of the XRR data was performed using the Leptos Reflectivity software [19].

To validate the multilayer stack model obtained from the XRR and further check the chemical species present, cross-sectional lamellas were investigated. They were prepared using an FEI Helios Nanolab 450 S focused ion beam and characterized by STEM and EDX. The bright-field images were acquired using a double spherical aberration-corrected JEOL JEM-ARM200F microscope operated at 200 kV. During lamella preparation, an ion beam source with currents varying from 7 pA to 0.23 nA at 30 keV was used to thin down the lamella. Final lamella thinning was carried out using a 7 pA ion beam current at 5 keV.

The EDX line profiles were performed using a liquid-nitrogen-free silicon drift detector.

3. Results

The results of this study are organized into different sections.

To start off, in Section 3.1, we perform a structural assessment of a typical projected PCM layer stack order [18] based on 10 nm Sb film. We focus on the PL-to-PC-material interface quality and stability, obtained by confining thin Sb film in between different layers. Several deposition approaches as well as stack variations are proposed and compared one by one. The limit case of a 3 nm thick Sb layer is discussed in Section 3.1.1.

Subsequently, in Section 3.2, we extend the study and compare the findings to another thin PC-material, Sb_2Te_3.

Eventually, to highlight the advantages of the new proposed stack variations, a thermal stability test of the PL-to-PC-material interface is presented in Section 3.3 for both the Sb and Sb_2Te_3 case studies.

The various samples and corresponding layer stack configurations investigated in this work are listed in Table 2. The results of their structural characterizations are detailed in the following sections.

Table 2. Overview of the samples investigated in this study. The layer order is listed from top to bottom.

Sample ID	Layer Stack Order (Top to Bottom)	Deposition T
Stack A	SiO_2/10 nm Sb/6 nm MeN/SiO_2	RT
Stack A *	SiO_2/10 nm Sb/6 nm MeN/SiO_2	200 °C
Stack B	SiO_2/6 nm MeN/10 nm Sb/SiO_2	200 °C
Stack C	SiO_2/6 nm MeN/10 nm Sb/SiO_2	RT
Stack D	SiO_2/3 nm Sb/6 nm MeN/SiO_2	RT
Stack E	SiO_2/6 nm MeN/3 nm Sb/SiO_2	200 °C
Stack F	SiO_2/10 nm Sb_2Te_3/6 nm MeN/SiO_2	RT
Stack G	SiO_2/6 nm MeN/10 nm Sb_2Te_3/SiO_2	200 °C

3.1. Projection Phase-Change Memory Stack with Sb

Figure 1c shows the XRR pattern of layer *Stack A*, as depicted in Figure 1a,b. The PC-material is deposited on the PL, and all layers are deposited at room temperature (RT). The inset of Figure 1c shows the stack schematic with the nominal thickness (not to scale).

To obtain quantitative information on the various layer densities, thickness and the interfacial roughness, the XRR curves were analyzed by fitting a simulated curve based on a multilayer model to the measured data [19]. The fit (black line) is in good agreement with the experimental data points (in lilac) in Figure 1. The outcome of the analysis is reported in Table 3.

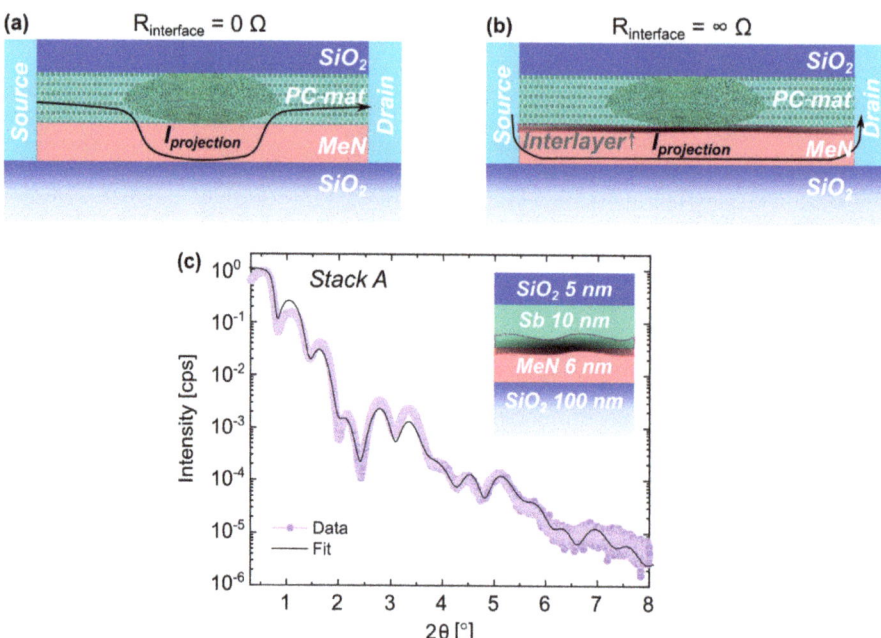

Figure 1. Sketch of a projected line-cell in cross sectional view with a partially amorphized PC-material layer: (**a**) in the scenario of "zero-interface resistance", the projection current mostly bypasses the amorphous phase; (**b**) if the interface resistance is infinite, the projection current bypasses the entire phase-change layer; (**c**) XRR scan (lilac) and fit comparison (black) for the original layer stack order, *Stack A*, used in previous projected phase-change memories [18]. The sample was deposited at RT.

Table 3. Output of the XRR fit analysis for *Stack A*.

Stack A	Thickness (nm)	Roughness (nm)	Density (g/cm^3)
SiO$_2$	5.6	0.3	2.3
c-Sb	5.2	0.3	6.2
a-Sb	4.4	0.1	4.5
Interlayer	0.8	0.3	1.0
MeN	4.4	0.3	13.5
SiO$_2$	100	0.3	2.4

The Sb film did not grow as a single uniform layer. An Sb layer with a lower density of 4.4 g/cm^3 was followed by a second one with a higher density of 6.2 g/cm^3. The latter agrees well with the density of crystalline Sb (c-Sb) found in the literature [20]. The lower density layer, instead, could be attributed to a residual amorphous onset layer (a-Sb). This was previously reported for the stabilization of thin (<10 nm) Sb and PC-materials stacked between different electrode materials [21–24].

A striking finding is the presence of an ultralow density (1.0 g/cm^3) layer between the liner and the PC-material that we refer to as a "spurious interfacial layer". It resulted in a poor adhesion between the Sb and the MeN, requiring a particularly careful handling of the samples during the fabrication of the actual line-cell devices.

The roughness at the various interfaces and at the surface was relatively smooth with values ≤0.3 nm (see Table 1). More details about the goodness of the fit and the significance of the extracted results are discussed in the Supplementary Materials (see Figure S1).

We propose several deposition variations to eliminate the "spurious interfacial layer" and to improve the interface stability between the different layers in the stack.

In a first approach, the deposition of all layers was performed at an elevated temperature (T = 200 °C), as it has been shown that temperature is a critical parameter to tune the growth of thin PC-materials [22–25]. The results of the structural investigation are shown in Figure 2a and Table 4. The deposition of the PC-material at elevated T improved the PL-to-PC-material interface. However, a thin onset of a 1 nm a-Sb layer with a density of 5.0 g/cm^3 was still needed for the fit.

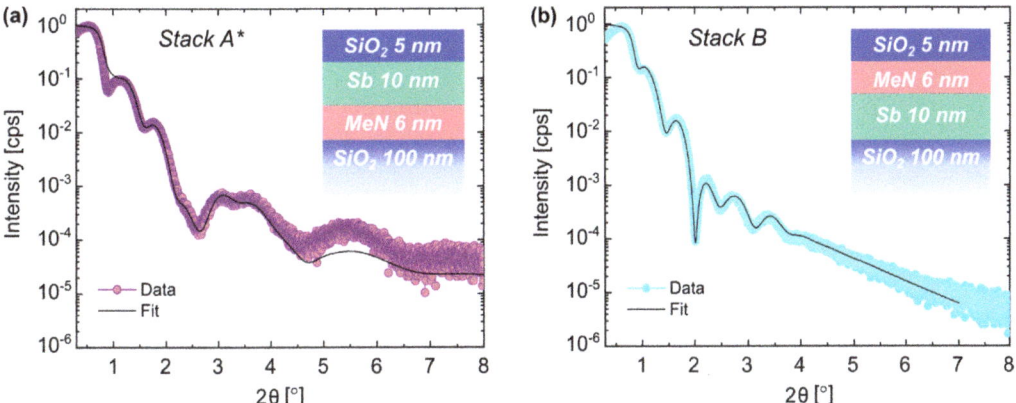

Figure 2. XRR scans and fit comparison for layer *Stack A **, deposited at T = 200 °C (**a**), and layer *Stack B* with inverted PC-material and projection layer order, deposited at T = 200 °C (**b**).

Table 4. Output of the XRR fit analysis for *Stack A **.

Stack A *	Thickness (nm)	Roughness (nm)	Density (g/cm^3)
SiO$_2$	5.3	1.0	2.2
c-Sb	8.4	0.7	5.6
a-Sb	1.0	0.5	5.0
MeN	4.2	0.3	13.8
SiO$_2$	100	0.1	2.4

The second approach was reordering the sequence of the thin films. That means the PC-material-to PL-order was inverted, as shown in the inset of Figure 2b. This path can be split into two alternatives. In one case, the layer deposition was performed at RT (see *Stack C* in the Supplementary Materials). In the other case, the layers were deposited at T = 200 °C (see Figure 2b). The advantages of inverting the order between these two layers are manifold. As shown in Table 5 and in Figure S2 of the Supplementary Materials, this approach eliminated the "spurious interfacial layer" between the PC-material and the MeN. Additionally, the deposition at elevated temperatures also improved the adhesion to the SiO$_2$ substrate (Table 5 vs. Table S1). A slight increase in the interfacial and surface roughness was obtained compared to the 0.3 nm value found for the RT depositions. Nevertheless, they were still reasonably smooth (≤1 nm) (Table 5 vs. Tables 3 and S1).

Table 5. Output of the XRR fit analysis for *Stack B*.

Stack B	Thickness (nm)	Roughness (nm)	Density (g/cm^3)
SiO$_2$	5.0	1.0	2.3
MeN	4.2	0.9	14.5
c-Sb	9.2	0.9	5.7
Interlayer	0.9	0.3	8.6
SiO$_2$	100	0.1	2.4

Next, we examined the microstructural properties of the films using XRD. We acquired and compared the XRD scans for *Stack A* (lilac scan), *Stack A ** (purple scan) and *Stack B* (cyan scan) in Figure 3. The peaks labeled as Sb (0003) and Sb (0006) correspond to the third and sixth order reflection of the PC-material film in the trigonal phase (R-3m space group) [26]. We made the following observations based on the characterization. There were differences in the peak areas between the traces of the three-layer stacks, which indicate that the crystalline Sb fractions in the thin films were different. It was the lowest for *Stack A* (deposited at RT) and higher for *Stack A ** and *Stack B*, which were both deposited at T = 200 °C. The presence of fringes in *Stack A ** highlights that the sample had high-quality interfaces with well-defined contrasts between the various layers in the stack. From the fringes' periodicity around the peak Sb (00-3) of *Stack B* (see inset of Figure 3), we extracted the thickness of the crystalline Sb layer, which amounted to 8.5 nm. This value compares well with the thickness obtained using XRR (see Table 4).

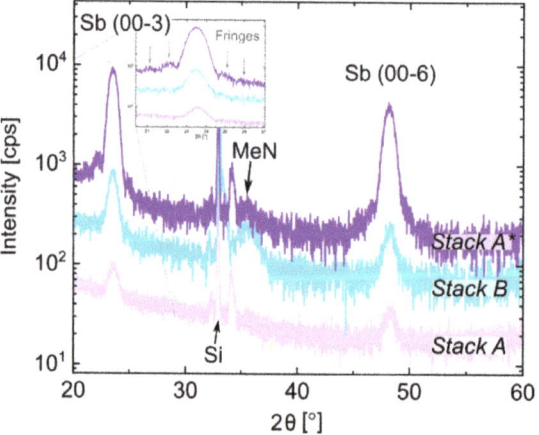

Figure 3. XRD profile comparison between *Stack A*, deposited at RT (lilac); *Stack A **, deposited at 200 °C (purple); and *Stack B*, deposited at 200 °C with an inverted PC-material/PL layer order (cyan). The fringes around the Sb (00-3) peak are highlighted in the inset.

The XRD profiles of the samples deposited at a high T show that the MeN layer started to crystallize. This is demonstrated by the broad band between 33° and 38° that is attributed to the MeN film. The sharp peaks in the same band are due to optics and the partial Si reflection from the substrate [27].

The deposition at elevated T and consequent crystallization of the MeN changed the resistivity of the layer with respect to the one that was intentionally chosen during the design of the device. The sheet resistance dependence of the MeN can be found in Figure S3. Considerations on the impact at the device level will follow in Section 4.

3.1.1. Case Study of Ultrathin Sb

When designing a PCM cell, downscaling the thickness of the PC-material is known to be an important parameter to tune device metrics, including the resistance window, programming currents and amorphous phase retention [16,24,28]. Hence, we investigated the limit case of an ultrathin 3 nm Sb (*Stack D*).

Figure 4 shows a STEM micrograph taken of a PCM cell based on *Stack D* (see schematic). Despite the layers originally being deposited at RT, the stack saw 180 °C during the fabrication of the cell.

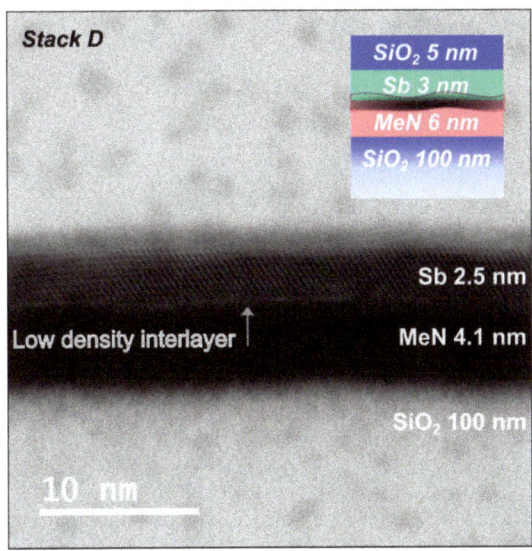

Figure 4. Cross-sectional bright-field STEM micrograph of *Stack D*, deposited at RT after device fabrication.

All layers can be clearly identified, and the PC-layer was mostly crystalline. Nevertheless, the bright thin layer at the interface between Sb and MeN indicates the presence of a low-density "spurious interfacial layer". If we compare this with the XRR analysis of the layer stack prior to device fabrication (see Table 6 and Figure S4), the results are in good agreement. Underneath the 1.7 nm crystalline Sb layer, a low-density (3.5 g/cm^3) interlayer was found. This layer, as found in *Stack A*, was also deposited at RT but with thicker Sb (10 nm) indicates poor adhesion and could potentially lead to delamination. The slight mismatch in thickness of the crystalline layer could be attributed to the elevated T seen by the layers during fabrication or to the known slight mismatch in thickness calibration between the STEM and XRR techniques. Eventually, the RT deposition of the projected PC-layer stack with ultrathin (3 nm) Sb led to similar results as the one with 10 nm Sb.

Table 6. Output of the XRR fit analysis for *Stack D* prior to PCM fabrication.

Stack D	Thickness (nm)	Roughness (nm)	Density (g/cm^3)
SiO$_2$	5.3	0.9	2.5
c-Sb	1.7	0.4	7.0
Interlayer	1.8	0.7	3.5
MeN	4.3	0.3	14.4
SiO$_2$	100	0.1	2.4

To further prove the advantages of the approach using elevated-T deposition and inverted PC-material and PL order, we applied it to the ultrathin 3 nm Sb case. The resulting sample was labeled as *Stack E*. The XRR profile and fit of *Stack E* are discussed based on Figure 5.

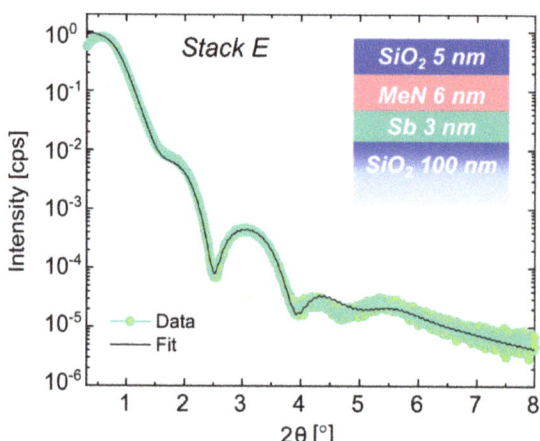

Figure 5. XRR scan and fit comparison for layer *Stack E*, deposited at T = 200 °C.

The decay of oscillations at higher angles (2θ > 4°) indicates a higher surface roughness with respect to the RT growth (see Figures 1 and 5). As obtained for the similarly grown *Stack B* with a thicker 10 nm PC-layer (see Figure 2b), there was no "spurious interlayer" at the interface between the now ultrathin PC-material and the PL nor between the PC-material and the bottom SiO$_2$ (see Table 7). On the contrary, the adhesion between the MeN and PC-material layers was improved, as indicated by the presence of the denser ultrathin Sb film between them. As for *Stack B*, also deposited at an elevated T, the roughness of the MeN and the SiO$_2$ layers increased to values of 0.7 and 0.8 nm, respectively, due to the temperature. The Sb layer was fully crystalline, i.e., no sign of a-Sb was noted.

Table 7. Output of the XRR fit analysis for *Stack E*.

Stack E	Thickness (nm)	Roughness (nm)	Density (g/cm^3)
SiO$_2$	4.6	0.8	2.0
MeN	4.9	0.7	14.4
c-Sb	2.4	0.5	6.2
Interlayer	0.5	0.1	8.1
SiO$_2$	100	0.1	2.4

3.2. Projection Phase-Change Memory Stack with Sb$_2$Te$_3$

With the intent of generalizing the results on Sb-based stacks to another confined thin PC-material system, the study presented in Section 3.1 was repeated on Sb$_2$Te$_3$.

Figure 6 shows the XRR curves of *Stack F* and *Stack G*, in blue and green, respectively. The former was deposited at RT with the original layer order, seeing the PC-material on top of the projection one. The latter was deposited at 200 °C with the inverted order of PC-layer and MeN. The results of the quantitative analysis derived from the fits (black lines) are listed in Tables 8 and 9, respectively. As for the samples with Sb, the decay in the oscillations at high angles (2θ > 4°) for deposition at elevated T was reflected by an increased interfacial and surface roughness. A validation of the XRR surface roughness for *Stack G* is given by atomic force microscopy in Figure S5.

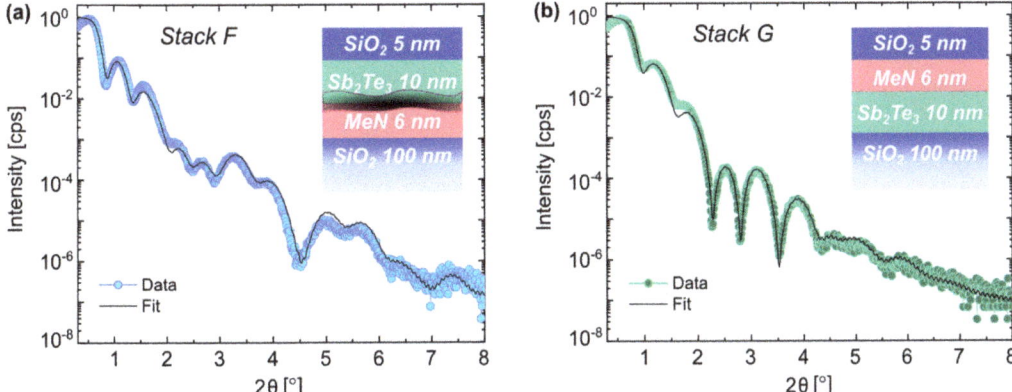

Figure 6. XRR scans and fit comparison for layer *Stack F*, deposited at RT (**a**), and layer *Stack G* with inverted PC-material and projection order. The sample was deposited at 200 °C (**b**).

Table 8. Output of the XRR fit analysis for *Stack F*.

Stack F	Thickness (nm)	Roughness (nm)	Density (g/cm³)
SiO$_2$	7.2	1.0	1.5
a-Sb$_2$Te$_3$	10	0.5	4.7
Interlayer	0.8	0.6	0.9
MeN	4	0.4	11.0
SiO$_2$	100	0.5	2.4

Table 9. Output of the XRR fit analysis for *Stack G*.

Stack G	Thickness (nm)	Roughness (nm)	Density (g/cm³)
SiO$_2$	6.1	0.7	1.0
MeN	3.9	0.8	12.3
2-Sb$_2$Te$_3$	4.9	0.7	5.7
1-Sb$_2$Te$_3$	3.9	0.6	6.3
SiO$_2$	100	0.3	2.4

By growing the PC-material on top of the MeN, an ultrathin layer of ultralow density was formed between the layers, which later resulted in delamination of the samples upon handling. The density of the PC-material was low, 4.7 g/cm³ on average, pointing toward a mostly amorphous layer. Instead, in *Stack G* (Figure 6b) the Sb$_2$Te$_3$ had a higher density, hinting at its crystallinity [29,30]. Despite being fully crystalline, a double layer (1-Sb$_2$Te$_3$ and 2-Sb$_2$Te$_3$) was needed to account for the density variations across the layer. The reason could be attributed to the presence of different strain conditions at the top and bottom interfaces with SiO$_2$ and MeN, resulting in inhomogeneous PC-material nucleation and crystallization across the layer [21,31].

To further validate our multilayer model of the two samples, STEM imaging was performed. The micrographs of *Stack F* and *Stack G* are shown in Figure 7a,b, respectively.

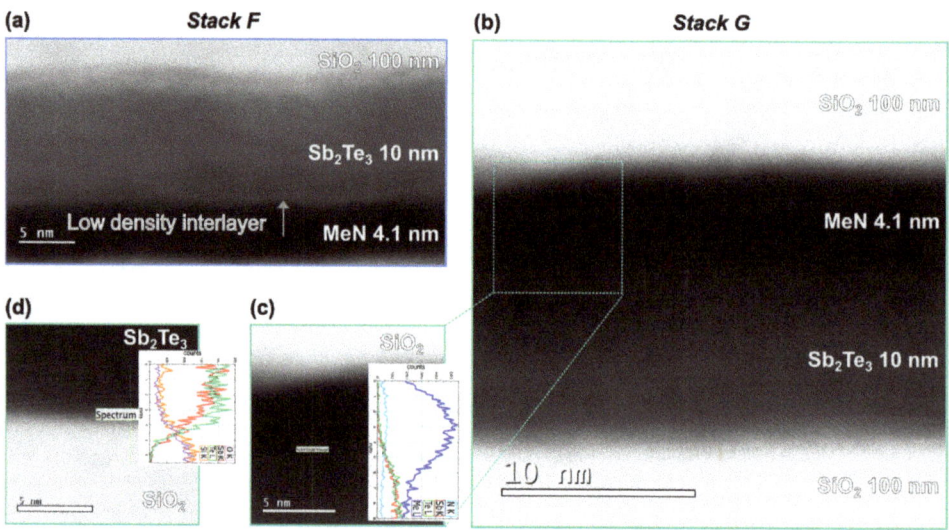

Figure 7. Cross-sectional bright-field STEM micrographs of *Stack F* (**a**), *Stack G* (**b**) and examples of zoomed-in images with line profiles on the MeN/Sb$_2$Te$_3$ interface in *Stack G* (**c**) and the Sb$_2$Te$_3$/SiO$_2$ interface of *Stack G* (**d**).

Figure 7a shows a STEM micrograph taken of a PCM cell based on *Stack F*. Despite the layers originally being deposited at RT, the stack saw an elevated T of 200 °C during the fabrication of the cell. All layers were clearly identified, and their thickness compares well with the one obtained from the XRR analysis. The Sb$_2$Te$_3$ layer appears predominantly crystalline, despite the lower density value (4.7 g/cm^3) obtained for *Stack F* prior to fabrication, which implied a mostly amorphous a-Sb$_2$Te$_3$ layer (see Table 8). The crystallinity of the layer in Figure 7a was attributed to the elevated T of 200 °C seen during fabrication.

A bright ultrathin layer was visible at the interface between the PC-material and the MeN. It corresponds to the "spurious interfacial layer", which was also detected in the XRR (Table 8).

The micrograph of *Stack G* shows distinct layers of thin films, and their overall thicknesses compare well with the ones obtained from the corresponding XRR (Table 9).

The Sb$_2$Te$_3$ layer was fully crystalline. Figure 7c is a typical zoomed-in image of the area near the PC-material/projection layer region to highlight the structural quality of this interface and the absence of any spurious low-density layers. Despite this, the EDX line scan detected a ~3 nm overlap of the metal (Me) and Sb/Te element signals across this interface. The grains of both Sb$_2$Te$_3$ and MeN could be responsible for this projection effect during the measurement. Moreover, Me injection into the Sb$_2$Te$_3$ lattice during the lamella preparation cannot be fully excluded. Moreover, in a previous study it was shown that the chemical affinity between two metals and the resulting Ti$_3$Te$_4$ compound formation at a higher T could explain the improved adhesion between the electrode and the PC-material [25]. Figure 7d is a zoomed-in image near the PC-material/SiO$_2$ region. The interface was sharp, and no "spurious interlayer" could be identified by the STEM or by the EDX analysis, as shown in the inset.

3.3. Thermal Stability Study of the Interfaces with Sb and Sb$_2$Te$_3$

To highlight the advantages of the newly proposed stack variation, a thermal stability test of the PL-to-PC-material interface is presented for both the Sb and the Sb$_2$Te$_3$ case studies. Figure 8 compares the XRR patterns of the RT-deposited *Stack A* (colored in lilac), with Sb as PC-material, and the one obtained after annealing at T = 220 °C for 30 min (in orange). The periods of the Pendellösung interference fringes are labeled as $\Delta\theta_1$ and $\Delta\theta_2$ for *Stack A* before and after annealing, respectively. The oscillations of *Stack A* before annealing (in lilac) display a slightly shorter oscillation period ($\Delta\theta_1 = 2.9° < \Delta\theta_2 = 3.0°$). Hence, a change in film thickness for one or more layers occurred upon annealing [32,33]. For larger scan angles ($2\theta > 4°$), the differences in the oscillation profile variations become more evident, highlighting major changes in the deeper layers of the stack. The curves showed no relevant decay of oscillation amplitude, suggesting that the interface roughness did not increase significantly. A change in surface roughness can be disregarded as well, as no fast decay of the XRR signal was detectable [33].

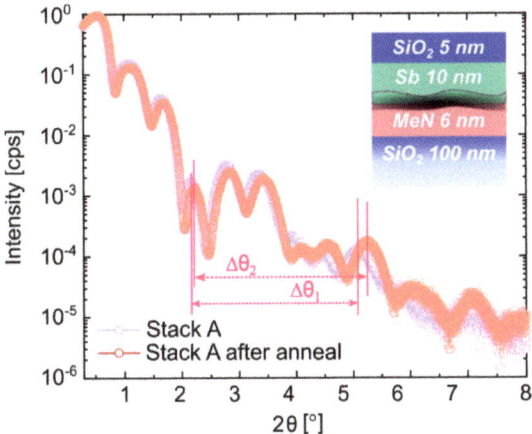

Figure 8. XRR scan comparison for layer *Stack A*, as deposited at RT (lilac), and after annealing at 220 °C (orange).

The result of the quantitative analysis of *Stack A* after annealing, obtained by fitting the measured data, is reported in Table 10 and confirms the assumptions based on the XRR profile comparisons. A slight increase (~0.7 nm) in the crystalline portion of Sb was detected compared to the sample before annealing, with an accompanying decrease in the amorphous layer contribution of ~1 nm (see Table 3 for comparison). The critical role of the "spurious interlayer" increased upon annealing at an elevated T, as demonstrated by its even lower density value of 0.3 g/cm^3 compared to 1.0 g/cm^3 for *Stack A* before annealing. This later resulted in a delaminated sample upon handling (not shown). See Tables 3 and 10 for comparisons.

Table 10. Output of the XRR fit analysis for *Stack A* after annealing at 220 °C.

Stack A (After Anneal.)	Thickness (nm)	Roughness (nm)	Density (g/cm^3)
SiO$_2$	5.7	0.1	1.9
c-Sb	5.9	0.1	6.3
a-Sb	3.3	0.3	4.6
Interlayer	1.1	0.1	0.3
MeN	4.2	0.3	12.6
SiO$_2$	100	0.1	2.4

The same temperature study on the Sb$_2$Te$_3$-based *Stack G*, with inverted PC-material and PL order, led to different results. Figure 9 compares the XRR profiles of *Stack G* before and after annealing at T = 220 °C for 30 min.

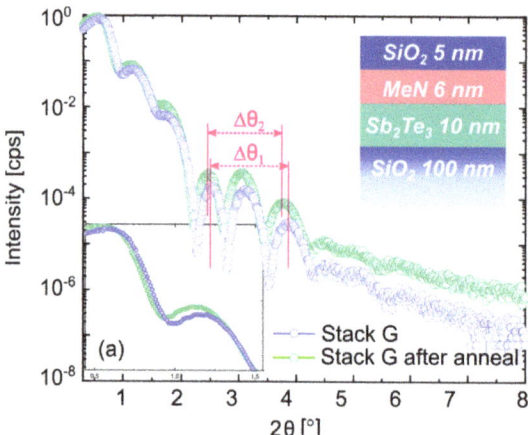

Figure 9. XRR scan comparison for layer *Stack G*, as deposited at 200 °C (blue) and after annealing at 220 °C (green). The Pendellösung fringes and their periodicity are marked with $\Delta\theta_1$ and $\Delta\theta_2$ (in red). (a) The inset shows a zoomed-in image of the total reflection edge.

In this case, the two curves show more pronounced differences. The total reflection edge is enlarged for the sake of clarity and shown in the inset (a) of Figure 9. The edge of the annealed film shifted toward smaller angles, indicating that one or more layers had a lower density with respect to the case before annealing [32,33]. The periods of the interference fringes are labeled as $\Delta\theta_1$ and $\Delta\theta_2$ for *Stack G* before and after annealing, respectively. From their comparison, information about the thickness change upon annealing can be extracted [33].

In particular, the oscillations of *Stack G* after annealing (green data points) displayed a shorter oscillation period ($\Delta\theta_2 = 1.3° < \Delta\theta_1 = 1.4°$) compared to before annealing (blue data points). Thus, an increase in film thickness occurred upon annealing. These observations were confirmed by the quantitative analysis and the change in the roughness, thickness and density of *Stack G* after annealing, as reported in Table 11 (see Table 9 for comparisons).

Table 11. Output of the XRR fit analysis for *Stack G* after annealing at 220 °C.

Stack G (After Anneal.)	Thickness (nm)	Roughness (nm)	Density (g/cm^3)
SiO$_2$	5.1	0.6	1.0
MeN	3.8	0.6	12.3
2-Sb$_2$Te$_3$	8.3	0.5	5.0
1-Sb$_2$Te$_3$	0.9	0.5	5.9
SiO$_2$	100	0.2	2.4

The temperature treatment converted the double Sb$_2$Te$_3$ layer into a uniform 8.3 nm layer of Sb$_2$Te$_3$ with a density of 5.0 g/cm^3. This value was lower compared to the ones reported in Table 9.

No ultralow-density "spurious interlayer" that was responsible for adhesion issues was detected at the interface with the SiO$_2$ or the interface with the projection layer, as desired. The curves show no relevant decay of the oscillation amplitude, suggesting that the interface roughness did not change significantly. Nevertheless, the slight decrease in the top surface layer roughness could be responsible for the higher XRR signal for *Stack G* after annealing.

4. Discussion

The results shown in Section 3 demonstrate how critical it is to control the layer stack deposition processes prior to projected phase-change memory fabrication. To fine-tune the layer and interfacial properties of the various materials in the stack, a solid methodology to characterize and model the multilayer stack is paramount.

Based on XRR, XRD, STEM and EDX, we have observed that by depositing the PC-material on the PL at RT, as performed in a previous study [18], a spurious interlayer is formed between the two layers. As a result of the ultralow density of the spurious layer, adhesion issues and, eventually, delamination can occur upon sample manipulation. In Kersting et al. [18], it was shown that in the limiting case of infinite interface resistance, the PCM cell could lose the multistate resistance property and eventually act as binary memory by completely bypassing the presence of the PL. The existence of such a "spurious layer" in the samples *Stack A*, *Stack D* and *Stack F* results in a non-ideal contact resistance between the PC-material and the MeN. We assume it to be a contributing factor, if not the main cause, of the non-ideal projection properties observed in the phase-change memory cells. These stacks lead to cells described by the scenario in Figure 1b.

Figure 10 shows an overview of the deposition approaches we proposed as well as stack variations to enhance the PC-material-to-PL interface and the uniformity of the PC-material.

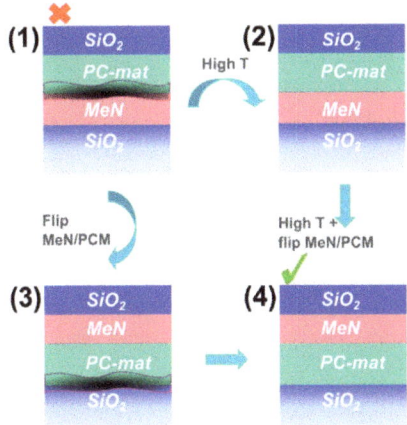

Figure 10. Summary of the four major stack deposition approaches pursued to enhance the PC-material-to-MeN interface and the uniformity of the PC-material.

(1) is the schematic of the original stack used for previous studies [18] that shows the spurious interface and results in easy delamination. In (2), the same layer order is kept, but the full stack is deposited at 200 °C. This leads to an improved PC-material-to-MeN interface, but nevertheless, the PC-material is not very uniform and, once processed into line-cell devices, it results in a binary behavior of the memory (not shown). In (3) the order between PC-material and MeN is inverted, but the full layer stack is deposited at RT. The advantage in this case is that there is no spurious layer between the PC-material and MeN, which is desirable. Nonetheless, a low-density interlayer is found between the PC-material and the bottom SiO_2 layer. The most robust approach is shown in (4). It combines approaches (2) and (3), yielding a sharp PC-material-to-projection-layer interface and a more uniform phase of the PC-layer. The adhesion of the PC-material to the bottom SiO_2 is also improved.

One more advantage of approach (4) is the larger freedom in designing the phase-change memory by decoupling the PC material deposition from that of the PL. The PC-material layer could be deposited at elevated T to improve the adhesion with the layer underneath and its phase uniformity. The PL on top could then be deposited at RT or any other T to consciously tune its resistivity. This precaution is important since properties such as the resistance of the PL have been shown to change with T (see Figure S2), leading to projected PCM devices that unintentionally differ from their original designs.

5. Conclusions

Projected PCM devices are an unfolding path toward the implementation of high-precision in-memory computing. However, to exploit their full potential, the origin of the interfacial resistance must be understood and controlled.

In this work, a dedicated structural investigation of the memory stacks using XRD, XRR, STEM and EDX revealed the presence of a spurious ultralow-density layer between the PC-material and the projection liner, which resulted in poor adhesion. This layer impaired the contact resistance between the PC-material and PL, leading to the suboptimal projection properties observed in the PCM cells.

To fine-tune the layers and interfacial properties of the various materials as well as to increase their stability in the memory cells, alternative deposition routes and stack designs were proposed and analyzed one by one.

We also established a schematic framework for designing projected phase-change memories for analog in-memory computing by advancing both the understanding and the tailoring of the interfacial resistance. The learning also applies to other applications where PCM devices are based on heterostructures.

Supplementary Materials: The following are available online at https://www.mdpi.com/article/10.3390/nano12101702/s1. Figure S1: XRR scan (lilac dots) and various fit comparisons for layer *Stack A*, deposited at RT; Figure S2: XRR scan (black) and fit comparison (red) for layer *Stack C*, deposited at RT and with inverted PC material/PL order. The various roughnesses, thicknesses and densities extracted from the fit are reported in the corresponding Table S1; Table S1: Output of the XRR fit analysis for *Stack C*; Figure S3: Temperature dependence of sheet resistance for MeN deposited with different nitrogen concentrations. Red data points refer to a RT deposition and green data points refer to a deposition performed at 200 °C; Figure S4: XRR scans and fit comparison for layer *Stack D*, deposited at RT; Figure S5: AFM image of a 5×5 µm^2 region on *Stack G*. The values of the measured roughness (Rq) are reported on the side.

Author Contributions: Conceptualization, A.S., V.B., S.G.S. and V.P.J.; methodology, V.B., S.G.S. and V.P.J.; software, V.B.; validation, V.B., V.P.J. and M.S.; formal analysis, V.B.; data curation, V.B., V.P.J. and M.S.; writing—original draft preparation, V.B.; writing—review and editing, V.P.J., M.S., S.G.S. and B.K.; funding acquisition, A.S. All authors have read and agreed to the published version of the manuscript.

Funding: This work was supported in part by the European Research Council through the European Union's Horizon 2020 Research and Innovation Program under grants 682675 and 966764.

Institutional Review Board Statement: Not applicable.

Informed Consent Statement: Not applicable.

Data Availability Statement: The data that support the findings of this study are available from the corresponding author upon reasonable request.

Acknowledgments: We thank the Cleanroom Operations Team of the Binnig and Rohrer Nanotechnology Center (BRNC) for their help and support and Mattia Halter for carefully reading the manuscript. We further acknowledge Linda Rudin for proofreading the manuscript.

Conflicts of Interest: The authors declare no conflict of interest. The funders had no role in the design of the study; in the collection, analyses, or interpretation of data; in the writing of the manuscript or in the decision to publish the results.

References

1. Raoux, S.; Wełnic, W.; Ielmini, D. Phase change materials and their application to nonvolatile memories. *Chem. Rev.* **2010**, *110*, 240–267. [CrossRef] [PubMed]
2. Wuttig, M.; Yamada, N. Phase-change materials for rewriteable data storage. *Nat. Mater.* **2007**, *6*, 824–832. [CrossRef] [PubMed]
3. Intel, Revolutionizing Memory and Storage. Available online: https://www.intel.in/content/www/in/en/design/products-and-solutions/memory-and-storage/pmem/revolutionizing-memory-and-storage-video.html (accessed on 29 March 2022).
4. Ambrogio, S.; Narayanan, P.; Tsai, H.; Shelby, R.M.; Boybat, I.; di Nolfo, C.; Sidler, S.; Giordano, M.; Bodini, M.; Farinha, N.C.P.; et al. Equivalent-accuracy accelerated neural-network training using analogue memory. *Nature* **2018**, *556*, 60–67. [CrossRef] [PubMed]
5. Ielmini, D.; Wong, H.-S.P. In-memory computing with resistive switching devices. *Nat. Electron.* **2018**, *1*, 333. [CrossRef]
6. Sebastian, A.; Le Gallo, M.; Khaddam-Aljameh, R.; Eleftheriou, E. Memory devices and applications for in-memory computing. *Nat. Nanotechnol.* **2020**, *15*, 529. [CrossRef]
7. Yu, S. Neuro-inspired computing with emerging nonvolatile memorys. *Proc. IEEE* **2018**, *106*, 260. [CrossRef]
8. Wong, H.-S.P.; Raoux, S.; Kim, S.; Liang, J.; Reifenberg, J.P.; Rajendran, B.; Asheghi, M.; Goodson, K.E. Phase Change Memory. *Proc. IEEE* **2010**, *98*, 2201. [CrossRef]
9. Sarwat, S.G.; Gehring, P.; Rodriguez Hernandez, G.; Warner, J.H.; Briggs, G.A.D.; Mol, J.A.; Bhaskaran, H. Scaling Limits of Graphene Nanoelectrodes. *Nano Lett.* **2017**, *17*, 3688. [CrossRef]
10. Raty, J.Y.; Zhang, W.; Luckas, J.; Chen, C.; Mazzarello, R.; Bichara, C.; Wuttig, M. Aging mechanisms in amorphous phase-change materials. *Nat. Commun.* **2015**, *6*, 1–8. [CrossRef]
11. Fantini, P.; Pirovano, A.; Ventrice, D.; Redaelli, A. Experimental investigation of transport properties in chalcogenide materials through noise measurements. *Appl. Phys. Lett.* **2006**, *88*, 263506. [CrossRef]
12. Ding, K.; Wang, J.; Zhou, Y.; Tian, H.; Lu, L.; Mazzarello, R.; Jia, C.; Zhang, W.; Rao, F.; Ma, E. Phase-change heterostructure enables ultralow noise and drift for memory operation. *Science* **2019**, *366*, 210. [CrossRef] [PubMed]
13. Koelmans Wabe, W.; Sebastian, A.; Jonnalagadda, V.P.; Krebs, D.; Dellmann, L.; Eleftheriou, E. Projected phase-change memory devices. *Nat. Commun.* **2015**, *6*, 8181. [CrossRef] [PubMed]
14. Kim, S.; Sosa, N.; BrightSky, M.; Mori, D.; Kim, W.; Zhu, Y.; Suu, K.; Lam, C. A Phase Change Memory Cell with Metallic Surfactant Layer as a Resistance Drift Stabilizer. In Proceedings of the 2013 IEEE International Electron Devices Meeting (IEDM), Washington, DC, USA, 9–11 December 2013; pp. 762–765.
15. Redaelli, A.; Pellizzer, F.; Pirovano, A. Phase Change Memory Device for Multibit Storage. European Patent No. EP2034536B1, 17 November 2010.
16. Sarwat, S.G.; Philip, T.M.; Chen, C.T.; Kersting, B.; Bruce, R.L.; Cheng, C.W.; Li, N.; Saulnier, N.; BrightSky, M.; Sebastian, A. Projected Mushroom Type Phase-Change Memory. *Adv. Fun. Mat.* **2021**, *31*, 2106547. [CrossRef]
17. Giannopoulos, I.; Sebastian, A.; Le Gallo, M.; Jonnalagadda, V.; Sousa, M.; Boon, M.N.; Eleftheriou, E. 8-bit Precision In-Memory Multiplication with Projected Phase-Change Memory. In Proceedings of the 2018 IEEE International Electron Devices Meeting (IEDM), San Francisco, CA, USA, 1–5 December 2018; pp. 27.7.1–27.7.4.
18. Kersting, B.; Ovuka, V.; Jonnalagadda, V.P.; Sousa, M.; Bragaglia, V.; Sarwat, S.G.; Le Gallo, M.; Salinga, M.; Sebastian, A. State dependence and temporal evolution of resistance in projected phase change memory. *Sci. Rep.* **2020**, *10*, 8248. [CrossRef] [PubMed]
19. Ulyanenkov, A. LEPTOS: A universal software for x-ray reflectivity and diffraction. *Proc. Adv. Comput. Methods X-ray NeutronOpt.* **2004**, *5536*, 1–15.
20. Persson, K. *Materials Data on Sb (SG:166) by Materials Project*; U.S. Department of Energy, Office of Scientific and Technical Information: Oak Ridge, TN, USA, 2015. [CrossRef]
21. Raoux, S.; Jordan-Sweet, J.L.; Kellock, A.J. Crystallization properties of ultrathin phase change films. *J. Appl. Phys.* **2008**, *103*, 114310. [CrossRef]
22. Raoux, S.; Cheng, H.-Y.; Jordan-Sweet, J.L.; Munoz, B.; Hitzbleck, M. Influence of interfaces and doping on the crystallization temperature of Ge-Sb. *Appl. Phys. Lett.* **2009**, *94*, 183144. [CrossRef]
23. Salinga, M.; Kersting, B.; Ronneberger, I.; Jonnalagadda, V.P.; Vu, X.T.; Le Gallo, M.; Giannopoulos, I.; Cojocaru-Mirédin, O.; Mazzarello, R.; Sebastian, A. Monatomic Phase Change Memory. *Nat. Mater.* **2018**, *17*, 681–685. [CrossRef]
24. Dragoni, D.; Behler, J.; Bernasconi, M. Mechanism of amorphous phase stabilization in ultrathin films of monoatomic phase change material. *Nanoscale* **2021**, *13*, 16146–16155. [CrossRef]
25. Loubriat, S.; Muyard, D.; Fillot, F.; Roule, A.; Veillerot, M.; Barnes, J.P.; Gergaud, P.; Vandroux, L.; Verdier, M.; Maitrejean, S. GeTe phase change material and Ti based electrode: Study of thermal stability and adhesion. *Microelectron. Eng.* **2011**, *88*, 817–821. [CrossRef]
26. Barrett, C.S.; Cucka, P.; Haefner, K. The crystal structure of antimony at 4.2, 78 and 298 K. *Acta Cryst.* **1963**, *16*, 451–453. [CrossRef]
27. Zaumseil, P. High-resolution characterization of the forbidden Si 200 and Si 222 reflections. *J. Appl. Cryst.* **2015**, *48*, 528–532. [CrossRef] [PubMed]
28. Jiao, F.; Chen, B.; Ding, K.; Li, K.; Wang, L.; Zeng, X.; Feng, R. Monatomic 2D phase-change memory for precise neuromorphic Computing. *Appl. Mat. Today* **2020**, *20*, 100641. [CrossRef]

29. Wang, R.; Calarco, R.; Arciprete, F.; Bragaglia, V. Epitaxial growth of GeTe/Sb$_2$Te$_3$ superlattices. *Mater. Sci. Semicond.* **2021**, *13*, 7106244. [CrossRef]
30. Greenwood, N.N.; Earnshaw, A. *Chemistry of the Elements*, 2nd ed.; Butterworth-Heinemann: Oxford, UK, 1997; pp. 581–582.
31. Guo, Q.; Li, M.; Li, Y.; Shi, L.; Chong Chong, T.; Kalb, J.A.; Thompson, C.V. Crystallization-induced stress in thin phase change films of different thicknesses. *Phys. Lett.* **2008**, *93*, 221907. [CrossRef]
32. Bragaglia, V.; Jenichen, B.; Giussani, A.; Perumal, K.; Riechert, H.; Calarco, R. Structural change upon annealing of amorphous GeSbTe grown on Si(111). *J. Appl. Phys.* **2014**, *116*, 054913. [CrossRef]
33. Birkholz, M. *Thin Film Analysis by X-ray Scattering*, 1st ed.; WILEY-VCH Verlag GmbH & Co.KGaA: Weinheim, Germany, 2006.

Article

Interface Formation during the Growth of Phase Change Material Heterostructures Based on Ge-Rich Ge-Sb-Te Alloys

Caroline Chèze [1], Flavia Righi Riva [1,*], Giulia Di Bella [1,2], Ernesto Placidi [2], Simone Prili [1], Marco Bertelli [3], Adriano Diaz Fattorini [3], Massimo Longo [3], Raffaella Calarco [3], Marco Bernasconi [4], Omar Abou El Kheir [4] and Fabrizio Arciprete [1]

1   Dipartimento di Fisica, Università di Roma "Tor Vergata", Via della Ricerca Scientifica 1, 00133 Rome, Italy; cheze_caroline@yahoo.fr (C.C.); giuliadibella05@gmail.com (G.D.B.); simone.prili@roma2.infn.it (S.P.); fabrizio.arciprete@roma2.infn.it (F.A.)
2   Department of Physics, Sapienza University of Rome, Piazzale Aldo Moro 5, 00185 Rome, Italy; ernesto.placidi@uniroma1.it
3   Istituto per la Microelettronica e Microsistemi (IMM), Consiglio Nazionale delle Ricerche (CNR), Via del Fosso del Cavaliere 100, 00133 Rome, Italy; marco.bertelli@artov.imm.cnr.it (M.B.); adriano.diazfattorini@artov.imm.cnr.it (A.D.F.); massimo.longo@artov.imm.cnr.it (M.L.); raffaella.calarco@artov.imm.cnr.it (R.C.)
4   Department of Materials Science, University of Milano-Bicocca, Via R. Cozzi 55, 20125 Milan, Italy; marco.bernasconi@unimib.it (M.B.); o.abouelkheir@campus.unimib.it (O.A.E.K.)
*   Correspondence: flavia.righiriva@roma2.infn.it

Abstract: In this study, we present a full characterization of the electronic properties of phase change material (PCM) double-layered heterostructures deposited on silicon substrates. Thin films of amorphous Ge-rich Ge-Sb-Te (GGST) alloys were grown by physical vapor deposition on Sb_2Te_3 and on $Ge_2Sb_2Te_5$ layers. The two heterostructures were characterized in situ by X-ray and ultraviolet photoemission spectroscopies (XPS and UPS) during the formation of the interface between the first and the second layer (top GGST film). The evolution of the composition across the heterostructure interface and information on interdiffusion were obtained. We found that, for both cases, the final composition of the GGST layer was close to Ge_2SbTe_2 (GST212), which is a thermodynamically favorable off-stoichiometry GeSbTe alloy in the Sb-GeTe pseudobinary of the ternary phase diagram. Density functional theory calculations allowed us to calculate the density of states for the valence band of the amorphous phase of GST212, which was in good agreement with the experimental valence bands measured in situ by UPS. The same heterostructures were characterized by X-ray diffraction as a function of the annealing temperature. Differences in the crystallization process are discussed on the basis of the photoemission results.

Keywords: PCM; Ge-rich Ge-Sb-Te alloys; heterostructures; electronic properties

1. Introduction

Phase change material (PCM) devices based on chalcogenide alloys constitute a well-established technology for optical storage and non-volatile electronic memories [1,2]. Moreover, PCMs are currently attracting increasing interest as promising candidates for neuromorphic applications [3]. The reason for such growing attention can be found in the requirements of modern information technology, which relies on the implementation of a large network of electronic smart systems, also known as the Internet of Things (IoT). The IoT can connect people, places, and systems, with a potential huge impact on several aspects of everyday life [4]. The impressive amount of data and information generated by modern electronic systems still requires large numbers of fast, cheap, and power-efficient embedded non-volatile memories and processing devices. A particular field of IoT is that of automotive applications, which must comply with rather strict requirements for the embedded electronic devices since they are expected to work at a temperature of 165 °C for at least 10 years, without any data loss. PCM-based memories have been found to be a suitable technology for the realization of rewritable storage media also in the automotive field. The ternary $(GeTe)_m(Sb_2Te_3)_n$ alloys are widely used as active material for the fabrication of both optical and electrical memories, commonly in the composition $Ge_2Sb_2Te_5$ (GST225) [5]. However, the GST225 crystallization temperature T_x of 150 °C is not sufficient to cope with the automotive requirements. Therefore, very recently, intense investigation into thermally stable PCMs with higher T_x has emerged [6,7]. GeSbTe with Ge-rich compositions (GGST) was found to be one of the most promising alloys [8]. However, thermally stable PCMs, such as GGST, suffer from a high write latency due to their low crystallization speed [9], and, in order to find a good compromise between stability and crystallization speed, new strategies have to be pursued. Investigations on superlattice structures made of alternating Sb_2Te_3 and GeTe or GST films [10–13] have revealed not only higher speed than GST, but also better thermal stability than Sb_2Te_3 films. This suggests that combining layers of materials with different physical properties can help to achieve the best material trade-off. Furthermore, investigations [10,13] have shown that the interfaces between the films in the superlattice, due to interdiffusion and alloying, can strongly affect the functional properties of the materials and consequently the overall performance of the final devices.

Here, following this strategy, we present a study on double-layer heterostructures based on GGST (high T_x but low crystallization speed) and Sb_2Te_3 or $Ge_2Sb_2Te_5$ (higher crystallization speed). To obtain insights on the properties of the interface during its formation, the structures were grown by successive depositions of the two PCM films. The electronic properties of the samples were studied in situ by X-ray and ultraviolet photoemission spectroscopies (XPS and UPS), with a particular focus on the formation of the interface to reveal the presence of intermixing and/or interdiffusion phenomena possibly occurring between the two layers. Moreover, the amorphous (a-) to crystalline (x-) phase transition was investigated by means of ex situ X-ray diffraction (XRD) under annealing conditions at increasing temperatures, revealing different crystallization behaviors between the two heterostructures.

2. Experiment

2.1. Sample Growth

The PCM films were deposited by Physical Vapor Deposition in a custom-made ultra-high vacuum (UHV) chamber system equipped with four Knudsen cells (Dr. Eberl MBE-Komponenten GmbH, Weil der Stadt, Germany) for the thermal evaporation of In (Azelis Electronics, Paris, France), Te, Sb and Ge (Alfa Aisar, Haverhill, MA, USA). The growth chamber was UHV-connected to the analysis chamber, allowing the in situ photoemission characterization of the samples. The two double-layered heterostructures investigated consisted of two PCM films of the same thickness deposited on $Si(100)/SiO_2$ substrates. In order to study the formation of the interface during the growth of the heterostructure, after the deposition of the first PCM layers (24 nm of Sb_2Te_3 or GST225), we carried out successive partial depositions of GGST, for resulting thicknesses of 1, 2, 4, 8, 16, and 24 nm (sample A: $Si(100)/SiO_2/Sb_2Te_3(24\ nm)/GGST(24\ nm)$; sample B: $Si(100)/SiO_2/GST225(24\ nm)/GGST(24\ nm)$; see scheme in Figure 1), and characterized each partial deposition by XPS and UPS. The films were grown using stoichiometric flux ratios. The Sb_2Te_3 layer was grown at 200 °C—and therefore in its crystalline phase—while all the others were grown nominally in the a-phase by keeping the substrate at room temperature. During the growth, the substrate temperature was monitored by a thermocouple positioned close to the sample holder and the pressure was in the range of high 10^{-9} Torr. In the case of sample B, after the XPS/UPS characterization, and before the GGST layer depositions, the a-GST225 layer was annealed in UHV at 300 °C to promote crystallization.

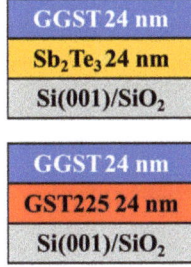

Figure 1. Schematics of the double-layer heterostructures.

2.2. Photoemission Characterization

For both sample A and B, photoemission experiments were performed in situ on the first PCM layer and after each partial deposition of the GGST layers.

XPS was carried out by using an Omicron DAR 400 Al/Mg Kα non-monochromatized X-ray (Taunusstein, Germany) source and a 100 mm hemispherical VG-CLAM2 electron spectrometer (Uckfield, UK). For all the XPS experiments, different core levels were considered: Te 3d, Sb 3d, Te 4d, Sb 4d, and Ge 3d acquired with the Mg Kα line, and Ge $2p_{3/2}$ acquired with the Al Kα line. Binding energies were calibrated by setting the Au $4f_{7/2}$ core level at 83.95 eV, measured on an Au foil in electrical contact with the sample. The collected XPS spectra were analyzed and fitted by the KolXPD software (version 1.8.0) and libraries (http://kolxpd.com accessed on 8 August 2018) using Voigt functions and Shirley background.

UPS measurements were carried out by using a VG helium discharge source set to excite the He I photon line.

2.3. X-ray Diffraction

XRD measurements were performed ex situ on replicas of sample A and B grown completely amorphous and after the deposition of a 10-nm-thick protective Si_3N_4 capping layer. XRD was carried out by a Bruker D8 Discover diffractometer (Billerica, MA, USA) equipped with a Cu X-ray source and an Anton Paar DHS1100 dome-type heating stage for temperature measurements in N_2 atmosphere. For each sample, grazing incidence ω-2θ XRD scans were acquired at increasing annealing temperatures from 30 to 300 °C under nitrogen flow. Annealing of sample A was performed with 10 steps of duration of 57 min each at a heating rate of 50 °C/min. Annealing of sample B was performed with 12 steps of duration of 25 min each at a heating rate of 60 °C/min.

2.4. Density Functional Theory Simulations

As will be shown in the next sections, the growth of samples A and B led to the formation of a top GGST layer very close to the GST212 composition.

Therefore, we computed the electronic density of states for a 216-atom model of amorphous GST212 generated by quenching from melt within molecular dynamics simulations based on the Born–Oppenheimer approximation and the solution of the electronic problem within Density Functional Theory (DFT). We used the exchange and correlation functional due to Perdew, Burke, and Ernzerhof (PBE) [14] within the same framework employed in our previous works on GST225 [15] and GGST [16] alloys, as detailed in the Supplementary Materials.

The structural properties were computed by averaging over a 12 ps trajectory at constant temperature and constant volume. The structural properties of our model of amorphous GST212 are similar to those reported in a previous DFT work [17]. Partial pair correlation functions, bond angle distribution functions, and the distribution of coordination numbers at 300 K are reported in Figures S1–S3 in the Supplementary Materials.

Average partial coordination numbers and the percentage of fraction of the different bonds are provided in Tables S1 and S2 in the Supplementary Materials, where we also report in Figure S4 the distribution of the q parameter, which allows the quantification of the fraction of Ge atoms in a tetrahedral bonding geometry. The distribution of rings is reported instead in Figure S5 in the Supplementary Materials. The structural parameters are overall in between those that we reported previously [16] for the two neighboring compositions, GST323 and GST423.

The electronic density of states (DOS) of the model optimized at zero temperature was computed from Kohn–Sham (KS) orbitals at the supercell Γ-point with the HSE06 hybrid functional [18] to better reproduce the band gap, as we have done in our previous works on GST225 [19,20]. KS energies were broadened by a Gaussian function with a variance of 80 meV.

3. Results and Discussion

A stack of Te 4d, Sb 4d, and Ge 3d XPS core levels as a function of the thickness of the deposited GGST layer is reported in Figure 2 for the grown heterostructures. Sb_2Te_3 was grown crystalline, while GST225 was grown amorphous and then annealed in UHV conditions at T = 300 °C before GGST deposition. Upon annealing (red curve in Figure 2c) all the core levels in the GST225 spectrum exhibited a chemical shift toward lower binding energy (BE) with respect to the as-grown a-GST225 (pink curve), being $\delta_{Te\,4d}$ = −0.32 eV, $\delta_{Sb\,4d}$ = −0.19 eV, and $\delta_{Ge\,3d}$ = −0.75 eV. This behavior reflects the change in chemical bonding upon the transition from a- to x-phase, with a lowering of the covalent character during crystallization. In the following, we will continue to refer to this film as x-GST225 (see also below the discussion of valence band measurements). After deposition of GGST, chemical shifts of the Te 4d and Ge 3d core levels toward higher BEs were clearly visible in the XPS spectra of both the heterostructures, while the opposite trend could be observed for Sb 4d (Figure 2b,c). To understand the behavior of the chemical shifts after the deposition of the GGST layers, it is useful to consider the expected BE shifts after the formation of binary compounds [21–23]. The Sb-Te bonding shifts the Sb 4d core levels toward higher BE (+0.6 eV with respect to metallic Sb) and the Te 4d toward lower BE (−0.55 eV with respect to metallic Te). Conversely, the Ge-Te bonding moves the Te 4d core levels of around 0.3 eV towards lower BE and the Ge 3d core levels to higher BE with a shift of +0.55 eV [21–23]. Therefore, after the formation of a ternary Ge-Sb-Te alloy, the BE of Te 4d core levels should be in the middle between the values expected for metallic Te and Sb_2Te_3. The more the alloy is Ge-rich, the more Te 4d core levels are shifted towards higher BEs. The same holds for Ge 3d core levels. In the case of Sb 4d, shifts towards lower BEs are expected. The BE shifts observed in our samples agree with this general trend if we consider the expected absolute BE positions for pure Te 4d, Sb 4d, and Ge 3d core levels (40.6, 32.1, and 29.3 eV, respectively). Moreover, at increasing GGST coverages, we observe a progressive increase in the shifts. To interpret this behavior, we need to consider the surface sensitivity of XPS, which limits the signal coming from the first layer at increasing thicknesses of the second layer. As a matter of fact, the collected spectra are the superposition of a decaying contribution from the first layer and an increasing one from the second layer, giving rise to an apparent progressive shift. The decay of the contribution of the first layer will be slower in the case of the formation of a rough interface. However, a similar behavior would also suggest the formation, at the interface, of an alloy with increasing Ge content. To discriminate the formation of alloys of varying composition due to a possible intermixing at the interface of the heterostructures, a quantitative analysis of XPS spectra is needed. It is important to note that the Te 4d, Sb 4d, and Ge 3d core levels fall within a narrow BE interval between 25 and 50 eV (see Figure 2), meaning that the kinetic energies of the collected photoelectrons are almost the same, as well as their escape depth. At these energies, the escape depth of the electrons is of the order of 2–3 nm [24], and the quantitative analysis of XPS spectra is therefore representative of the bulk composition of the samples. The analysis of the Te 4d, Sb 4d, and Ge 3d core levels is thus particularly important because, by a fit deconvolution of the experimental spectra (normalized for the sensitivity factor of the specific element [24]), we can determine, in situ, the composition of the alloys as a function of the GGST thickness.

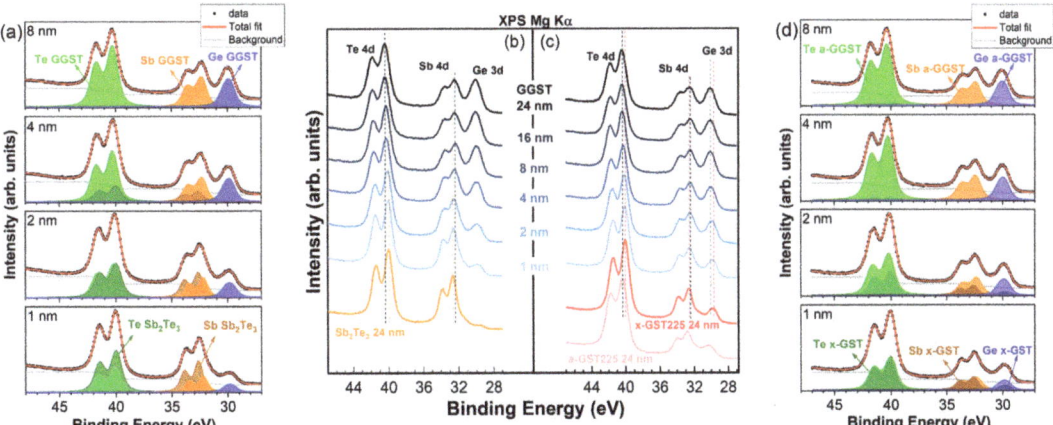

Figure 2. XPS spectra evolution of the Te 4d, Sb 4d, and Ge 3d core levels for (**b**) sample A, (**c**) sample B at increasing thicknesses of the GGST film. Chemical shifts toward higher and lower BEs are visible for Te 4d, Ge 3d, and Sb 4d, respectively. Fit deconvolution of the core levels for selected thicknesses (1, 2, 4, and 8 nm) of the deposited GGST layer is also shown for (**a**) sample A, (**d**) sample B.

For the best fit deconvolution of the spectra, we modeled the system by means of two components for each of the elements in the alloys, one representing the contribution from the first PCM layer and the second one representing the GGST second layer. The first layer will be considered unchanged while the second layer will result as the average of the growing second layer and the intermixed interface, if any. Therefore, the best fits were obtained considering a vanishing contribution from the bottom PCM layers (Sb_2Te_3 or GST225) with fixed BEs, Lorentzian and Gaussian widths, spin–orbit splits, and branching ratios of the Voigt doublets.

In Figure 2a,d, the obtained results for selected representative thicknesses (1, 2, 4, 8 nm) of the deposited GGST film are reported for sample A and B, respectively. Within this model, if there is no intermixing at the interface, the components associated with the GGST growing layer would result at their final BE already after the first partial deposition (1 nm), regardless of the morphology of the interface. Therefore, the observation of a progressive shift at increasing GGST coverages suggests, in our case, a varying composition of the growing layer due to intermixing occurring at the interface during the evolution of the heterostructure. Obviously, intermixing involves both sides of the interface, but in our model, the intermixed layer is averaged in the contribution associated with the second layer.

In Figure 3a,b, we plot the BE of the core levels for each element of the GGST growing layer for sample A and B, respectively, as determined from the fitting of the XPS spectra (light green, light orange, and blue curves in Figure 2a,d). The results clearly show that, for both heterostructures, the components associated with the second layer undergo a progressive shift in the first partial depositions, reaching a stable average value for GGST thicknesses higher than 4 nm. It is interesting to note that, while the final BE of the Ge and Te core levels are comparable for both heterostructures, some differences can be revealed in the behavior of the Sb core levels. After the deposition of 24 nm of GGST, the chemical shift toward lower BEs of Sb core levels is larger in sample A than in sample B. Moreover, after the very first depositions, for both heterostructures, the Sb 4d BE decreases toward the energy of metallic Sb and then increases again after 2 nm (a possible Sb segregation at the interface, not included in the model, cannot be ruled out). Again, this effect appears to be larger for sample A (see Figure 3a). These differences can be understood if we consider that, in heterostructure A, the first layer does not contain Ge, revealing the high sensitivity of the Sb 4d core levels to the Ge content.

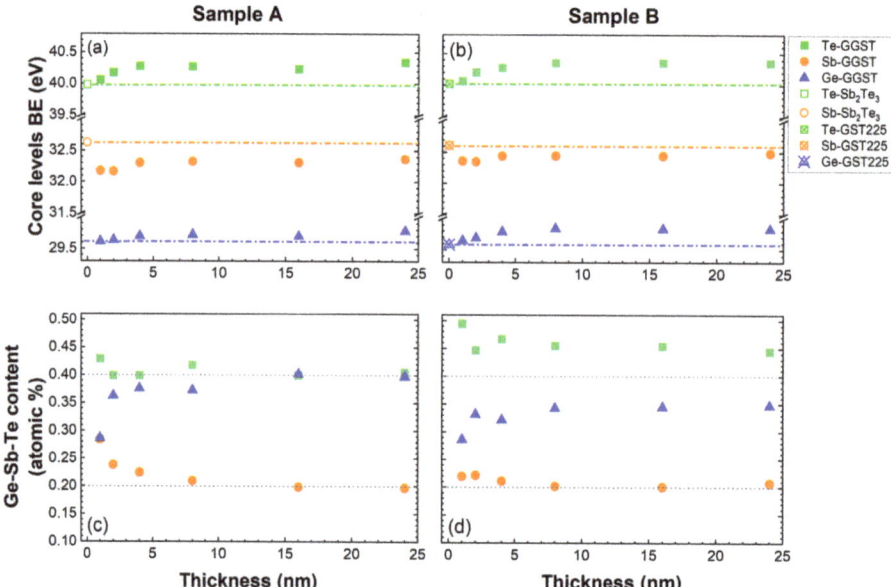

Figure 3. Core level BEs of the Sb_2Te_3 Te 4d and Sb 4d (open green square and orange circle, respectively), GST225 Te 4d, Sb 4d, Ge 3d (open crossed green square, orange circle, and blue triangle, respectively), and GGST Te 4d, Sb 4d, Ge 3d (green squares, orange circles, and blue triangles, respectively) for sample A (**a**) and sample B (**b**) as a function of the GGST layer thickness as estimated by fit deconvolution of the XPS spectra of Figure 1. The evolution of the composition of the GGST layer is shown for sample A (**c**) and sample B (**d**).

To obtain further insights on this behavior and on the interface formation, the stoichiometry as a function of the thickness of the second layer was determined from the fitting of the core levels. The results are summarized in Figure 3c,d, where the compositions in atomic percentage are reported. For sample A, starting from the Sb_2Te_3 layer, such atomic percentages are compatible with composition Sb:2.0, Te:2.6, and the GGST layer continuously enriches in Ge and Te, from approximately Ge:2.0, Sb:2.0, Te:3.0 up to the final composition Ge:4.0, Sb:2.0, Te:4.1. In the case of sample B, the GST225 layer with composition Ge:1.1, Sb:2.0, Te:6.3 in the a-phase becomes a crystalline mixed 124/225 (Ge:1.3, Sb:2.0, Te:4.6) after annealing; as a matter of fact, the 124 composition is the most stable among the alloys along the pseudo-binary line [25,26]. The composition of the GGST layer evolves to a final Ge:3.4, Sb:2.0, Te:4.5, a value rather close to 424 as for the case of Sb_2Te_3/GGST. Considering these results, the slightly larger BE variation observed for Sb core levels after the deposition of the 24-nm-thick GGST layer in sample A can be explained with the formation of a slightly Ge-richer final composition. For both the heterostructures, the composition of the GGST layer progressively changes (see Tables S5 and S6 in the Supplementary Materials), reaching a stable average value for thicknesses higher than approximately 8 and 4 nm for sample A and B, respectively. The observed evolution of composition is compatible with the occurrence of intermixing phenomena at the interface between the two PCM layers, a hypothesis also supported by the progressive chemical shifts of the GGST core levels. By comparing the results in Figure 3c,d and considering the probing depth of XPS at these energies (which is approximately 2–3 nm), we can suggest that the intermixed region is larger in sample A than in B.

As for the XPS experiments, UPS spectra of the heterostructures were collected during the formation of the interface after each deposition step of the GGST layer. The results are summarized in Figure 4, where a change in the UPS spectra line-shape is visible for both heterostructures already after the deposition of 1 nm of GGST, with the valence band maximum (VBM) progressively shifting towards higher BEs at increasing thicknesses (and increasing Ge content) of the GGST overlayer from 1 to 24 nm. The VB of the 24-nm-thick GGST has a mixed Te 5p, Sb 5p, and Ge 4p character in the region between 6 and 0 eV and, for both sample A and B, is characterized by a prominent peak at approximately 2 eV, and two broad features at approximately 3.5 and 5 eV. The same line-shape can be observed in the UPS spectrum of the as-grown a-GST225 layer of sample B (pink curve in Figure 3), with the VB showing the same three main features around 2.0, 3.5, and 5.0 eV, typical features of amorphous GST [27]. Upon annealing (red curve in Figure 4b), GST225 crystallizes and the VB shows a change in the line-shape with the formation of clear peaks at approximately 0.7, 1.8, and 3.0 eV of BE and the energy of the VBM shifting toward the Fermi level indicated by the green line in Figure 4. Such a shift can be ascribed to an increased hole concentration due to the formation of vacancies in the Ge/Sb sublattice [27–29]. Interestingly, the experimental spectra for annealed x-GST225 have very close correspondence with the DOS calculated by DFT for an ordered trigonal t-GST structure [29]. The a- to x-transition of GST225 is also visible in the XPS spectra in Figure 2c (red curves). These observations, associated with a mixed 124/225 composition, suggest that annealing in UHV at 300 °C is effective in triggering the transition of a-GST225 to the trigonal phase. Similar results for the VB have been found by Klein et al. [28] for cubic GST124, Caravati et. al. for cubic GST225 [19,29], and by Akola and Jones [30] for cubic Ge-rich GST8 2 11.

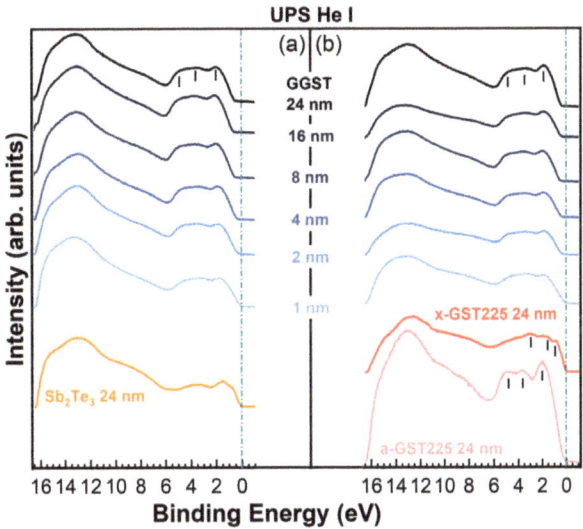

Figure 4. UPS spectra evolution measured for (**a**) sample A, (**b**) sample B as a function of the thickness of the deposited layers.

A good correspondence with simulations was found also in the case of sample A for the Sb_2Te_3 VB (orange curve in Figure 4), which compares well to the theoretical DOS of crystalline Sb_2Te_3 [31,32] and to the experimental VB measured in situ for a polycrystalline sample [33], showing a mixed Te-Sb p character in the region 0–6 eV and a dominant s character in the range 7–13 eV.

By inspection of the UPS spectra collected on the 24-nm-thick GGST films of both heterostructures, a shift of approximately 0.11 eV of the VBM towards higher BEs could be detected for sample A with respect to B, which is reasonable if we consider that sample B has a slightly lower Ge content, as determined by XPS [34]. It is also worth noting that the deposition of only 1 nm of GGST changes the VB line-shape of both samples, which shows the typical features of amorphous GST. The VBM shifts progressively as a function of GGST thickness up 8 nm and 2/4 nm for samples A and B, respectively, suggesting that the interface of sample B is sharper than the interface of sample A. Since the final composition of the GGST films obtained by XPS is close to 424 for both heterostructures, the VB measured by UPS for the 24 nm GGST layers was compared to the electronic DOS calculated by DFT simulations for amorphous GST212, as shown in Figure 5b. Experimental and theoretical spectra show quite good correspondence; we remark that previous DFT calculations [35] have revealed that alloys on the Sb-GeTe isoelectronic line (on average three p-electrons per atom) are particularly stable in the cubic crystalline phase. Their free energy is in fact very close to the convex hull built from the thermodynamically stable compositions [35]. This feature has resulted from the construction of a map of the formation free energy of GST alloys in the central part of the ternary phase diagram generated by high-throughput DFT calculations in Ref. [35]. We might expect that compositions on the Sb-GeTe line, such as GST212, are also more stable than others in the ternary phase diagram in the amorphous phase as well, although we do not have a map similar to that of Ref. [35] for the amorphous phase. Should this be the case, it would justify the formation of the GST212 composition after the growth of samples A and B reported above.

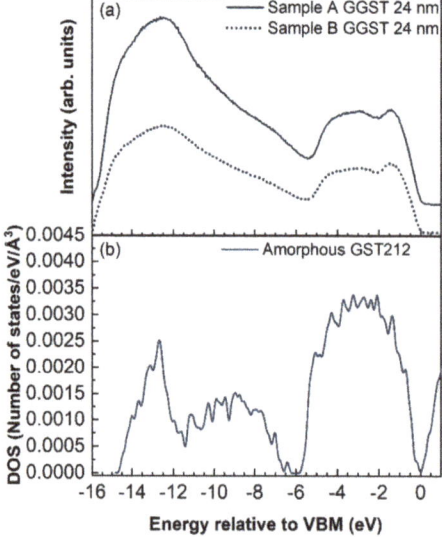

Figure 5. Comparison between (a) UPS spectra acquired on the two heterostructures after deposition of 24 nm of GGST and (b) DFT calculations of the DOS of amorphous GST212.

The a- to x-transition of the two heterostructures was investigated by ex situ XRD measurements performed at increasing temperatures on replicas of samples A and B grown completely amorphous. The diffractograms as a function of annealing temperature are reported in Figure 6 for sample A at the selected temperatures of 30, 200, and 300 °C (blue curves) and for sample B at 30, 150, and 300 °C (green curves). The XRD curves collected on single layers of Sb_2Te_3 (orange curve), cubic c-GST225 (red curve), and trigonal t-GST225 (violet curve) as reference diffractograms for cubic and trigonal polycrystalline GST are also shown. As shown in Figure 6, the diffractograms of the heterostructures do not display any peak at 30 °C, thus confirming that the samples were both grown amorphous. A first visible difference between the two samples is the crystallization temperature, which is lower for sample B ($T_x \sim 150$ °C) than for sample A ($T_x \sim 200$ °C), coherent with the fact that the GGST layer in this latter sample is richer in Ge. In the case of sample B annealed at 150 °C, four diffraction peaks can be seen at $2\theta = 25.7°, 29.6°, 42.7°, 69.7°$ that can be, respectively, identified as (111), (002), (022), (024) Bragg reflections of the c-GST. Interestingly, the contribution of the (111) and (024) reflections remains stable up to T = 350 °C (data not shown), while the other orientations disappear. In the case of sample A, the peak around 40°, identified as the (106) reflection of t-GST, is found together with the reflections at $2\theta = 25.7°, 29°, 42.5°, 69.6°$ of c-GST at 200 °C. Upon a further increase in the temperature up to 300 °C, the contributions of the c-GST(002) and t-GST(106) reflections increase in intensity, while the c-GST(022) reflection remains almost stable. Furthermore, a faint reflection from c-GST(222) can be identified at 52.2°. These observations suggest that the crystallization of the two heterostructures occurs with different behaviors: while a mixed cubic–trigonal phase at around 200 °C is found for sample A, sample B shows a transition to the cubic phase already at around 150 °C. The second layer of sample A is Ge-richer than sample B and this can justify a higher T_x, although a Tx > 250 °C would be expected for GGST [8]. This behavior is remarkable and suggests that the crystallization of the GGST can be affected by the presence of the underlayer. In particular, the introduction of a Sb_2Te_3 (sample A) or GST225 (sample B) underlayer beneath the GGST layer induces a lowering of the crystallization temperature in comparison with GGST single layers. Furthermore, although there are several reports that phase separation can occur upon annealing of GGST single layers, leading to the crystallization of Ge and other GST phases [9,36,37], no evidence of Ge segregation could be found in the XRD curves collected on our GGST-based heterostructures at increasing annealing temperature. All these observations suggest that the nature of the first layer and the extent of the interface can affect the crystallization behavior upon annealing. From the collected data, sample B results to be more stable at increasing annealing temperatures, which could be related to the fact that this system is characterized by a sharper interface than Sb_2Te_3/GGST, as suggested by the XPS and UPS results.

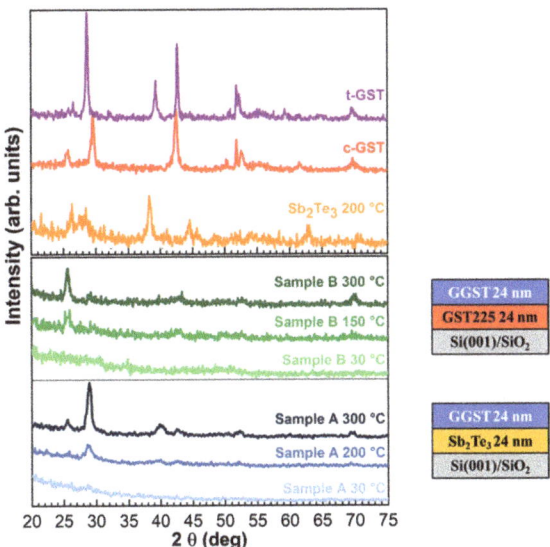

Figure 6. ω-2θ XRD curves as a function of the annealing temperature for sample A and B. Single layer (Sb_2Te_3, c-GST225, and t-GST225 used as reference for cubic and trigonal polycrystalline GST) XRD scans are shown as references.

4. Conclusions

In this work, we present a full photoemission characterization of the interface formation in two double-layer PCM heterostructures: Sb_2Te_3/GGST and GST225/GGST.

The evolution of the heterostructure was followed as a function of the thickness of the GGST film. Interdiffusion between the GGST layer and the underlayer was highlighted for both cases, and to a larger extent for Sb_2Te_3/GGST. At increasing thicknesses of the GGST layer, a progressive increase in the Ge and Te content was found, which led to a final composition close to GST212 for both heterostructures. Once again, from the XPS measurements, evidence of the crystallization of the first GST225 layer could be found for sample B upon annealing in UHV, leading to a mixed GST124/GST225 composition and to a clear change in the VB line-shape. UPS measurements performed on the a-GGST layers also revealed quite good agreement between the experimental spectra and the DOS of amorphous GST212 calculated by DFT.

From the XRD data recorded upon annealing of both heterostructures, different behavior during the crystallization of the two systems could be found, with sample B resulting in a more stable heterostructure at increasing annealing temperatures. Furthermore, XRD measurements suggested that Sb_2Te_3 can act as a template for the crystallization of the GGST layer, leading for sample A to the formation of a mixed cubic–trigonal phase, unlike sample B, which maintained a stable cubic phase. Considering these results, it will be of interest to study multiple PCM layers and investigate the effect of the layer thickness on the crystallization process [38,39].

Supplementary Materials: The following supporting information can be downloaded at: https://www.mdpi.com/article/10.3390/nano12061007/s1, Figure S1: Total and partial pair correlation functions of the DFT model of amorphous GST212 at 300 K. The dashed lines indicate the bonding cutoff used to define the coordination numbers, which is 3.2 Å for all pairs except for Sb-Te for which the bonding cutoff was set to 3.4 Å; Figure S2: Bond angle distribution function of the DFT model of amorphous GST212 at 300 K. The total distribution and the distributions resolved for the different types of the central atom are shown; Figure S3: Distribution of the coordination numbers of the DFT model of amorphous GST212 obtained by using the bonding cutoff defined by the partial pair correlation functions in Figure S1; Figure S4: Distribution of the local order parameter q for tetrahedricity resolved for atomic species and coordination number for amorphous GST212. To obtain a continuous distribution, the order parameter for individual atoms is broadened with a Gaussian function 0.005 wide. The local order parameter q was introduced in [8] and it is defined by $q = 1 - \frac{3}{8}\sum_{i<k}\left[\frac{1}{3} + \cos\theta_{ijk}\right]$ where the sum runs over the couples of atoms bonded to a central atom j and forming a bonding angle θ_{ijk}. The order parameter evaluates q = 1 for the ideal tetrahedral geometry and q = 5/8 for a four-coordinated defective octahedral site. As shown in our previous work [9], the integration of the q distribution of the 4-coordinated Ge atoms from 0.8 to 1.0 gives a measure of the fraction of Ge atoms tetrahedrally coordinated, which turns out to be 45.2% for GST212; Figure S5: Ring distribution function of amorphous GST212; Table S1: Average coordination numbers for different pairs of atoms computed from the partial pair corre-lation functions (Figure S1) for the DFT model of amorphous GST212; Table S2: Percentage fraction of the different types of bonds in the DFT model of amorphous GST212; Table S3: Binding energies, amplitudes, Lorentzian widths, Gaussian widths of the Te 4d, Sb 4d and Ge 3d core levels as a function of the deposited thickness for sample A; Table S4: Binding energies, amplitudes, Lorentzian widths, Gaussian widths of the Te 4d, Sb 4d and Ge 3d core levels as a function of the deposited thickness for sample B; Table S5: Evolution of the composition across heterostructure A; Table S6: Evolution of the composition across heterostructure B [14,16,40–46].

Author Contributions: Conceptualization, F.A.; data curation, C.C., F.R.R., G.D.B., E.P., M.B. (Marco Bertelli), A.D.F., M.L. and F.A.; formal analysis, M.B. (Marco Bernasconi) and O.A.E.K.; funding acquisition, R.C., M.B. (Marco Bernasconi) and F.A.; investigation, C.C., F.R.R., G.D.B., E.P., S.P., M.B. (Marco Bertelli), A.D.F., M.L., R.C., M.B. (Marco Bernasconi), O.A.E.K. and F.A.; methodology, M.B. (Marco Bernasconi) and F.A.; project administration, R.C.; software, O.A.E.K.; supervision, F.A.; writing—original draft, F.R.R.; writing—review & editing, C.C., F.R.R., G.D.B., E.P., S.P., M.B. (Marco

Bertelli), A.D.F., M.L., R.C., M.B. (Marco Bernasconi), O.A.E.K. and F.A. All authors have read and agreed to the published version of the manuscript.

Funding: This research was founded by EU Horizon 2020 GA No. 824957.

Data Availability Statement: The data that support the findings of this study are available from the corresponding author upon reasonable request.

Acknowledgments: This project has received funding from the European Union's Horizon 2020 research and innovation program under Grant Agreement No. 824957 ("BeforeHand:" Boosting Performance of Phase Change Devices by Hetero- and Nanostructure Material Design).

Conflicts of Interest: The authors declare no conflict of interest.

References

1. Burr, G.; Breitwisch, M.J.; Franceschini, M.; Garetto, D.; Gopalakrishnan, K.; Jackson, B.; Kurdi, B.; Lam, C.; Lastras, L.A.; Padilla, A.; et al. Phase Change Memory Technology. *J. Vac. Sci. Technol. B Microelectron. Nanom. Struct. Process. Meas. Phenom.* **2010**, *28*, 223–262. [CrossRef]
2. Lencer, D.; Salinga, M.; Wuttig, M. Design Rules for Phase-Change Materials in Data Storage Applications. *Adv. Mater.* **2011**, *23*, 2030–2058. [CrossRef] [PubMed]
3. Sebastian, A.; Le Gallo, M.; Khaddam-Aljameh, R.; Eleftheriou, E. Memory devices and applications for in-memory computing. *Nat. Nanotechnol.* **2020**, *15*, 529–544. [CrossRef]
4. Bandyopadhyay, D.; Sen, J. Internet of Things: Applications and Challenges in Technology and Standardization. *Wirel. Pers. Commun.* **2011**, *58*, 49–69. [CrossRef]
5. Wuttig, M.; Yamada, N. Phase-change materials for rewriteable data storage. *Nat. Mater.* **2007**, *6*, 824–832. [CrossRef] [PubMed]
6. Koch, C.; Hansen, A.L.; Dankwort, T.; Schienke, G.; Paulsen, M.; Meyer, D.; Wimmer, M.; Wuttig, M.; Kienle, L.; Bensch, W. Enhanced temperature stability and exceptionally high electrical contrast of selenium substituted $Ge_2Sb_2Te_5$ phase change materials. *RSC Adv.* **2017**, *7*, 17164–17172. [CrossRef]
7. Saxena, N.; Persch, C.; Wuttig, M.; Manivannan, A. Exploring ultrafast threshold switching in In_3SbTe_2 phase change memory devices. *Sci. Rep.* **2019**, *9*, 19251. [CrossRef]
8. Zuliani, P.; Palumbo, E.; Borghi, M.; Dalla Libera, G.; Annunziata, R. Engineering of chalcogenide materials for embedded applications of Phase Change Memory. *Solid. State. Electron.* **2015**, *111*, 27–31. [CrossRef]
9. Privitera, S.M.S.; López García, I.; Bongiorno, C.; Sousa, V.; Cyrille, M.C.; Navarro, G.; Sabbione, C.; Carria, E.; Rimini, E. Crystallization properties of melt-quenched Ge-rich GeSbTe thin films for phase change memory applications. *J. Appl. Phys.* **2020**, *128*, 155105. [CrossRef]
10. Cecchi, S.; Zallo, E.; Momand, J.; Wang, R.; Kooi, B.J.; Verheijen, M.A.; Calarco, R. Improved structural and electrical properties in native $Sb_2Te_3/Ge_xSb_2Te_{3+x}$ van der Waals superlattices due to intermixing mitigation. *APL Mater.* **2017**, *5*, 26107. [CrossRef]
11. Simpson, R.E.; Fons, P.; Kolobov, A.V.; Fukaya, T.; Krbal, M.; Yagi, T.; Tominaga, J. Interfacial phase-change memory. *Nat. Nanotechnol.* **2011**, *6*, 501–505. [CrossRef]
12. Boniardi, M.; Boschker, J.E.; Momand, J.; Kooi, B.J.; Redaelli, A.; Calarco, R. Evidence for Thermal-Based Transition in Super-Lattice Phase Change Memory. *Phys. Status Solidi Rapid Res. Lett.* **2019**, *13*, 1800634. [CrossRef]
13. Wang, R.; Calarco, R.; Arciprete, F.; Bragaglia, V. Epitaxial growth of $GeTe/Sb_2Te_3$ superlattices. *Mater. Sci. Semicond. Process.* **2022**, *137*, 106244. [CrossRef]
14. Perdew, J.P.; Burke, K.; Ernzerhof, M. Generalized Gradient Approximation Made Simple. *Phys. Rev. Lett.* **1996**, *77*, 3865–3868. [CrossRef] [PubMed]
15. Caravati, S.; Bernasconi, M.; Kühne, T.D.; Krack, M.; Parrinello, M. Coexistence of tetrahedral- and octahedral-like sites in amorphous phase change materials. *Appl. Phys. Lett.* **2007**, *91*, 171906. [CrossRef]
16. Abou El Kheir, O.; Dragoni, D.; Bernasconi, M. Density functional simulations of decomposition pathways of Ge-rich GeSbTe alloys for phase change memories. *Phys. Rev. Mater.* **2021**, *5*, 95004. [CrossRef]
17. Sun, L.; Zhou, Y.X.; Wang, X.D.; Chen, Y.H.; Deringer, V.L.; Mazzarello, R.; Zhang, W. Ab initio molecular dynamics and materials design for embedded phase-change memory. *NPJ Comput. Mater.* **2021**, *7*, 29. [CrossRef]
18. Krukau, A.V.; Vydrov, O.A.; Izmaylov, A.F.; Scuseria, G.E. Influence of the exchange screening parameter on the performance of screened hybrid functionals. *J. Chem. Phys.* **2006**, *125*, 224106. [CrossRef]
19. Caravati, S.; Bernasconi, M.; Kühne, T.D.; Krack, M.; Parrinello, M. First-principles study of crystalline and amorphous $Ge_2Sb_2Te_5$ and the effects of stoichiometric defects. *J. Phys. Condens. Matter.* **2009**, *21*, 255501. [CrossRef]
20. Cobelli, M.; Dragoni, D.; Caravati, S.; Bernasconi, M. Metal-semiconductor transition in the supercooled liquid phase of the $Ge_2Sb_2Te_5$ and GeTe compounds. *Phys. Rev. Mater.* **2021**, *5*, 45004. [CrossRef]
21. Sarkar, I.; Perumal, K.; Kulkarni, S.; Drube, W. Origin of electronic localization in metal-insulator transition of phase change materials. *Appl. Phys. Lett.* **2018**, *113*, 263502. [CrossRef]
22. Nolot, E.; Sabbione, C.; Pessoa, W.; Prazakova, L.; Navarro, G. Germanium, antimony, tellurium, their binary and ternary alloys and the impact of nitrogen: An X-ray photoelectron study. *Appl. Surf. Sci.* **2021**, *536*, 147703. [CrossRef]

23. Baeck, J.H.; Ann, Y.K.; Jeong, K.H.; Cho, M.H.; Ko, D.H.; Oh, J.H.; Jeong, H. Electronic structure of Te/Sb/Ge and Sb/Te/Ge multi layer films using photoelectron spectroscopy. *J. Am. Chem. Soc.* **2009**, *131*, 13634–13638. [CrossRef] [PubMed]
24. Briggs, D.; Seah, M.P. *Practical Surface Analysis*; Wiley: Hoboken, NJ, USA, 1992; Volume 1.
25. Bragaglia, V.; Jenichen, B.; Giussani, A.; Perumal, K.; Riechert, H.; Calarco, R. Structural change upon annealing of amorphous GeSbTe grown on Si(111). *J. Appl. Phys.* **2014**, *116*, 54913. [CrossRef]
26. Da Silva, J.; Walsh, A.; Lee, H. Insights into the structure of the stable and meta-stable (GeTe)$_m$(Sb$_2$Te$_3$)$_n$ compounds. *Phys. Rev. B* **2008**, *78*, 224111. [CrossRef]
27. Lee, D.; Lee, S.S.; Kim, W.; Hwang, C.; Hossain, M.B.; Hung, N.L.; Kim, H.; Kim, C.G.; Lee, H.; Hwang, H.N.; et al. Valence band structures of the phase change material Ge$_2$Sb$_2$Te$_5$. *Appl. Phys. Lett.* **2007**, *91*, 251901. [CrossRef]
28. Klein, A.; Dieker, H.; Späth, B.; Fons, P.; Kolobov, A.; Steimer, C.; Wuttig, M. Changes in electronic structure and chemical bonding upon crystallization of the phase change material GeSb$_2$Te$_4$. *Phys. Rev. Lett.* **2008**, *100*, 16402. [CrossRef]
29. Sosso, G.C.; Caravati, S.; Gatti, C.; Assoni, S.; Bernasconi, M. Vibrational properties of hexagonal Ge$_2$Sb$_2$Te$_5$ from first principles. *J. Phys. Condens. Matter.* **2009**, *21*, 245401. [CrossRef]
30. Akola, J.; Jones, R.O. Structure of amorphous Ge$_8$Sb$_2$Te$_{11}$: GeTe-Sb$_2$Te$_3$ alloys and optical storage. *Phys. Rev. B* **2009**, *79*, 134118. [CrossRef]
31. Caravati, S.; Bernasconi, M.; Parrinello, M. First-principles study of liquid and amorphous Sb$_2$Te$_3$. *Phys. Rev. B* **2010**, *81*, 014201. [CrossRef]
32. Sosso, G.C.; Caravati, S.; Bernasconi, M. Vibrational properties of crystalline Sb$_2$Te$_3$ from first principles. *J. Phys. Condens. Matter* **2009**, *21*, 95410. [CrossRef] [PubMed]
33. Bendt, G.; Zastrow, S.; Nielsch, K.; Mandal, P.S.; Sánchez-Barriga, J.; Rader, O.; Schulz, S. Deposition of topological insulator Sb$_2$Te$_3$ films by an MOCVD process. *J. Mater. Chem. A* **2014**, *2*, 8215–8222. [CrossRef]
34. Kim, J.J.; Kobayashi, K.; Ikenaga, E.; Kobata, M.; Ueda, S.; Matsunaga, T.; Kifune, K.; Kojima, R.; Yamada, N. Electronic structure of amorphous and crystalline (GeTe)$_{1−x}$(Sb2 Te3)$_x$ investigated using hard x-ray photoemission spectroscopy. *Phys. Rev. B—Condens. Matter Mater. Phys.* **2007**, *76*, 115124. [CrossRef]
35. Abou El Kheir, O.; Bernasconi, M. High-Throughput Calculations on the Decomposition Reactions of Off-Stoichiometry GeSbTe Alloys for Embedded Memories. *Nanomaterials* **2021**, *11*, 2382. [CrossRef]
36. Di Biagio, F.; Cecchi, S.; Arciprete, F.; Calarco, R. Crystallization Study of Ge-Rich (GeTe)$_m$(Sb$_2$Te$_3$)$_n$ Using Two-Step Annealing Process. *Phys. Status Solidi—Rapid Res. Lett.* **2019**, *13*, 1800632. [CrossRef]
37. Agati, M.; Vallet, M.; Joulié, S.; Benoit, D.; Claverie, A. Chemical phase segregation during the crystallization of Ge-rich GeSbTe alloys. *J. Mater. Chem. C* **2019**, *7*, 8720–8729. [CrossRef]
38. Wang, R.; Bragaglia, V.; Boschker, J.E.; Calarco, R. Intermixing during Epitaxial Growth of van der Waals Bonded Nominal GeTe/Sb$_2$Te$_3$ Superlattices. *Cryst. Growth Des.* **2016**, *16*, 3596–3601. [CrossRef]
39. Momand, J.; Wang, R.; Boschker, J.E.; Verheijen, M.A.; Calarco, R.; Kooi, B.J. Interface formation of two- and three-dimensionally bonded materials in the case of GeTe-Sb$_2$Te$_3$ superlattices. *Nanoscale* **2015**, *7*, 19136–19143. [CrossRef]
40. Goedecker, S.; Teter, M.; Hutter, J. Separable dual-space Gaussian pseudopotentials. *Phys. Rev. B* **1996**, *54*, 1703–1710. [CrossRef]
41. Krack, M. Pseudopotentials for H to Kr optimized for gradient-corrected exchange-correlation functionals. *Theor. Chim. Acta* **2005**, *114*, 145–152. [CrossRef]
42. VandeVondele, J.; Krack, M.; Mohamed, F.; Parrinello, M.; Chassaing, T.; Hutter, J. Quickstep: Fast and accurate density functional calculations using a mixed Gaussian and plane waves approach. *Comput. Phys. Commun.* **2005**, *167*, 103–128. [CrossRef]
43. Grimme, S.; Antony, J.; Ehrlich, S.; Krieg, H. A consistent and accurate ab initio parametrization of density functional dispersion correction (DFT-D) for the 94 elements H-Pu. *J. Chem. Phys.* **2010**, *132*, 154104–154119. [CrossRef] [PubMed]
44. Njoroge, W.K.; Wöltgens, H.-W.; Wuttig, M. Density changes upon crystallization of Ge2Sb2.04Te4.74 films. *J. Vac. Sci. Technol. A Vac. Surf. Film.* **2002**, *20*, 230–233. [CrossRef]
45. Errington, J.R.; DeBenedetti, P.G. Relationship between structural order and the anomalies of liquid water. *Nature* **2001**, *409*, 318–321. [CrossRef]
46. Sosso, G.C.; Caravati, S.; Mazzarello, R.; Bernasconi, M. Raman spectra of cubic and amorphous Ge2Sb2Te5 from first principles. *Phys. Rev. B* **2011**, *83*, 134201. [CrossRef]

Article

Interface Analysis of MOCVD Grown GeTe/Sb$_2$Te$_3$ and Ge-Rich Ge-Sb-Te/Sb$_2$Te$_3$ Core-Shell Nanowires

Arun Kumar [1,2,*], Seyed Ariana Mirshokraee [1,3], Alessio Lamperti [1], Matteo Cantoni [3], Massimo Longo [4,*] and Claudia Wiemer [1]

[1] CNR—Institute for Microelectronics and Microsystems, Via C. Olivetti 2, 20864 Agrate Brianza, Italy; s.mirshokraee@campus.unimib.it (S.A.M.); alessio.lamperti@mdm.imm.cnr.it (A.L.); claudia.wiemer@mdm.imm.cnr.it (C.W.)
[2] Department of Physics 'E.R. Caianiello', University of Salerno, Via G. Paollo I 132, 84084 Salerno, Italy
[3] Department of Physics, Politecnico di Milano, Via G. Colombo 81, 20133 Milano, Italy; matteo.cantoni@polimi.it
[4] CNR—Institute for Microelectronics and Microsystems, Via del Fosso del Cavaliere 100, 00133 Rome, Italy
* Correspondence: arun.kumar@mdm.imm.cnr.it (A.K.); massimo.longo@artov.imm.cnr.it (M.L.)

Abstract: Controlling material thickness and element interdiffusion at the interface is crucial for many applications of core-shell nanowires. Herein, we report the thickness-controlled and conformal growth of a Sb$_2$Te$_3$ shell over GeTe and Ge-rich Ge-Sb-Te core nanowires synthesized via metal-organic chemical vapor deposition (MOCVD), catalyzed by the Vapor–Liquid–Solid (VLS) mechanism. The thickness of the Sb$_2$Te$_3$ shell could be adjusted by controlling the growth time without altering the nanowire morphology. Scanning electron microscopy (SEM) and X-ray diffraction (XRD) techniques were employed to examine the surface morphology and the structure of the nanowires. The study aims to investigate the interdiffusion, intactness, as well as the oxidation state of the core-shell nanowires. Angle-resolved X-ray photoelectron spectroscopy (XPS) was applied to investigate the surface chemistry of the nanowires. No elemental interdiffusion between the GeTe, Ge-rich Ge-Sb-Te cores, and Sb$_2$Te$_3$ shell of the nanowires was revealed. Chemical bonding between the core and the shell was observed.

Keywords: MOCVD; XPS; Ge-rich Ge-Sb-Te/Sb$_2$Te$_3$; GeTe/Sb$_2$Te$_3$; core-shell nanowires

1. Introduction

Interest in Phase Change Memories (PCMs) based on chalcogenide alloys continues to grow due to the wide range of possible applications that can be reached by the reversible amorphous–to–crystalline phase transition of chalcogenide materials [1–6]. PCMs are the most promising candidate for realizing "Storage Class Memories", which could fill the gap between "operation" and "storage" memories [7,8]. The main improvements needed to exploit the full potential of PCMs in these innovative applications are the reduction of the programming currents and further cell downscaling. In particular, the reduction in the programming currents is related to lower energy to induce the phase transitions, hence lower power consumption. However, other limitations, such as alloy composition, structure, small cell size, and high scalability, exist in the path of the realization of PCMs [9]. Further, multilevel PCMs' heterostructures have been explored by properly pairing two or more phase change materials featuring different crystallization properties. Compared to the commonly employed Ge-Sb-Te (GST) and GeTe (GT) alloys in PCM cells, Sb$_2$Te$_3$ (ST) exhibits a lower crystallization temperature and faster reversible switching, making it an ideal candidate for the realization of heterostructure-based multi-level PCM cells, in which the different properties of the involved materials can be combined to improve the overall performances, and for carrying out fundamental studies on the role of the heterointerface in the resistive switch.

Among the different methods explored to overcome the above limitations, nanowire (NWs) based PCMs fabricated by the bottom-up approach have drawn considerable interest [10–15]. Advantages of using NWs for such an application are their small sub lithographic feature sizes and single-crystalline defect-free structure, where novel functionalities are expected to originate by engineering the constituent compositions, sizes, and structures, as in the case of axial [16–18], radial (core-shell) [19,20], and branched heterostructured NWs [21]. Various PCM cells based on the GT and GST structures have been previously reported [22–28]. Core-shell NWs formed by two chalcogenide materials with different phase change characteristics, namely $Ge_2Sb_2Te_5/GeTe$, have also been proposed as multi-level PCM memory cells [29]. Indeed, the growth and the properties of Ge-rich Ge-Sb-Te/Sb_2Te_3 (GGST/ST) and GeTe/Sb_2Te_3 (GT/ST) PCM core-shell NWs have been recently reported by our group [30,31].

In the present work, we examined the morphological and structural characteristics of the GT/ST and GGST/ST core-shell NWs via Field Emission Scanning electron microscopy (FESEM) and X-ray diffraction (XRD), respectively. A detailed investigation was carried out on the elemental composition and chemical bonding of the MOCVD-grown NWs by exploiting X-ray photoelectron spectroscopy (XPS) analysis on both core and core-shell structures, to extract information on the existing nanointerfaces between the different core and shell materials.

2. Experimental Section

The growth of the GGST and GT core NWs was carried out with an Aixtron AIX200/4 MOCVD reactor (Aixtron SE, Herzogenrath, Germany), employing the Vapor–Liquid–Solid (VLS) mechanism catalyzed by Au nanoparticles (NPs), with average sizes of 20, 30, and 50 nm. Details about the growth methodology have been previously reported [30,31]. The GT core NWs were obtained with optimized reactor temperature (T), reactor pressure (P), and a growth duration (t) of 400 °C, 50 mbar, and 60 min, respectively. The required precursor pressure for the GT growth was 3.35×10^{-3} mbar for tetraisdimethylamino germanium ($Ge[N(CH_3)_2]_4$, TDMAGe), and 8.58×10^{-3} mbar for diisopropyl telluride (($C_3H_7)_2Te$, DiPTe). The GGST core NWs were obtained with the optimized T, P, and t parameters of 400 °C, 50 mbar, and 60 min, respectively. The required precursor pressures were 4.42×10^{-3} mbar for tetraisdimethylamino germanium ($Ge[N(CH_3)_2]_4$, TDMAGe), 5.12×10^{-5} mbar for antimony trichloride ($SbCl_3$) and 6.98×10^{-3} mbar for diisopropyl telluride (($C_3H_7)_2Te$, DiPTe). The chemical composition of the GGST NWs has been previously estimated by electron energy loss spectroscopy (EELS) to be 35% Ge, 10% Sb, 55% Te, corresponding to the atomic ratio of Ge:Sb:Te = 3:1:5, with the Ge concentration higher than regular $Ge_2Sb_2Te_5$ [30].

The ST shell deposition on the core NWs was performed at room temperature, by employing the $SbCl_3$ and bis(trimethylsilyl) telluride ($Te(SiMe_3)_2$, DSMTe) precursors, with the partial pressures of 2.23×10^{-4} mbar and 3.25×10^{-4} mbar, respectively. The growth rate of ST was optimized to be 20 nm/h.

The surface morphology in-plane and cross-section mode of the obtained NWs was carried out using a ZeissR© Supra40 field-emission scanning electron microscope (FE-SEM) (Cark Zeiss, Oberkochen, Germany). XRD analysis was performed using an Ital Structures HRD3000 diffractometer system (Ital Structures Sas, Riva de Garda, Italy) to evaluate the average crystal structure of the obtained NWs. The experimental XRD patterns were analyzed by the MAUD program. XPS was performed by an XPS ESCA 5600 apparatus (monochromatic Al K_α X-ray source, 1486.6 eV) equipped with a concentric hemispherical analyzer (Physical Electronics Inc., Chanhassen, MN, USA) to investigate the elemental composition and the chemical bonding of NWs, and the interface between the core and shell of the core-shell NWs. Pass energy was 58.50 eV with energy steps of 0.25 eV. Binding energies were calibrated considering the C_1s peak at 285 eV. All the related XPS measurements were analyzed and fitted by the XPSPEAK41 program, using a Voigt shape for each experimental peak.

3. Results and Discussion

Figure 1a,b shows the top-view SEM images of the GT and GGST core NWs grown on Si and SiO$_2$ substrates, respectively. It is useful to recall here the results of the morphological analysis from our previous works [30,31], Au catalyst NPs were observed at the NWs' tips, confirming that the growth occurs via the VLS mechanism; the as-grown GT NWs were of an average length and diameter of about 5 μm and 50 nm, while the GGST NWs were about 1.40 μm long and 60 nm in diameter, respectively; the diameter of the NWs turned out to be directly dependent on the size of the catalyzed Au NPs [30,31]. Figure 1c,d depicts the GT/ST and GGST/ST core-shell NWs having shells of about 10 nm thickness, and the insets show the magnified view of a single NW. The 10 nm shell revealed a lower granularity in comparison to the 30 nm shell [30,31]. The shell was continuously deposited all over the core NWs. Such nanostructures with 10 nm and 30 nm shell thicknesses (obtained for a deposition time of 30 and 90 min, respectively) were, therefore, the subject of the present study, with a special focus on the core-shell interface.

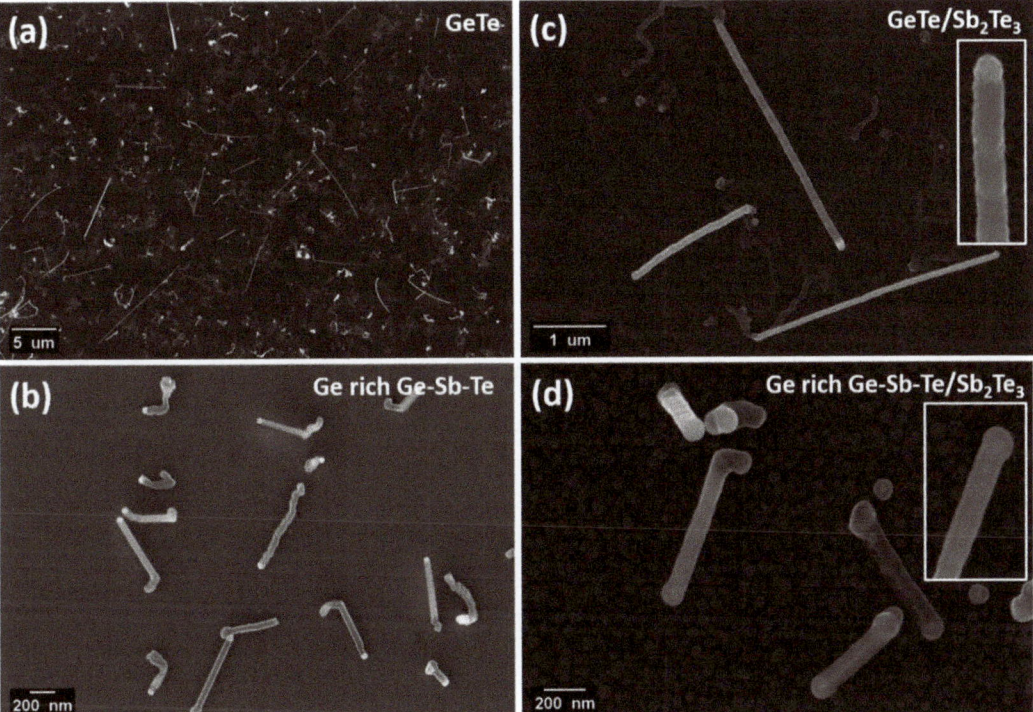

Figure 1. Top-view SEM image of (**a**) GT, (**b**) GGST core; and (**c**) GT/ST, (**d**) GGST/ST core-shell NWs with 10 nm shell thickness, insets show the magnified view of a single NW.

Figure 2 shows the XRD patterns obtained from the core and core-shell NWs. The patterns were simulated by taking into account not only the peak position but also the existing background and peak broadening, using the open-source software Maud [32]. Figure 2a shows a set of XRD patterns obtained for the GGST core and GGST/ST core-shell NWs. The GGST core NWs exhibited broad diffraction peaks centered at the 2θ values expected for the face-centered cubic Ge$_2$Sb$_2$Te$_5$ phase, along with the presence of the cubic Au NPs diffraction peak at about 38.1° (Figure 2a). Thus, the crystallized cubic GGST, exhibits the lattice parameters of cubic Ge$_2$Sb$_2$Te$_5$ [30]. In the case of GGST/ST core-shell NWs, the Ge$_2$Sb$_2$Te$_5$ peak is revealed as a shoulder close to the (015) reflection of ST, while the diffraction peaks from the Sb_2Te_3 phase are clearly visible (Figure 2a). Further, Figure 2b shows the XRD patterns of the GT and GT/ST core-shell NWs. The GT NWs were found to exhibit rhombohedral structure. The extracted lattice parameters of the NWs were found to be a = 8.28 Å, and c = 10.55 Å. In the pattern of the GT/ST core-shell NWs in Figure 2b, there is a shoulder at the right side of the (015) reflection of ST, at around 2θ = 29.6°, that could be attributed to the (202) main reflection of the GT structure, confirming that the core GT NWs preserve their crystallinity after the ST deposition. The extracted lattice parameters of Sb_2Te_3 by Rietveld refinements were found to be a = 4.22 Å and c = 30.46 Å [31]. This also confirms that no structural disordering occurred with the deposited shell.

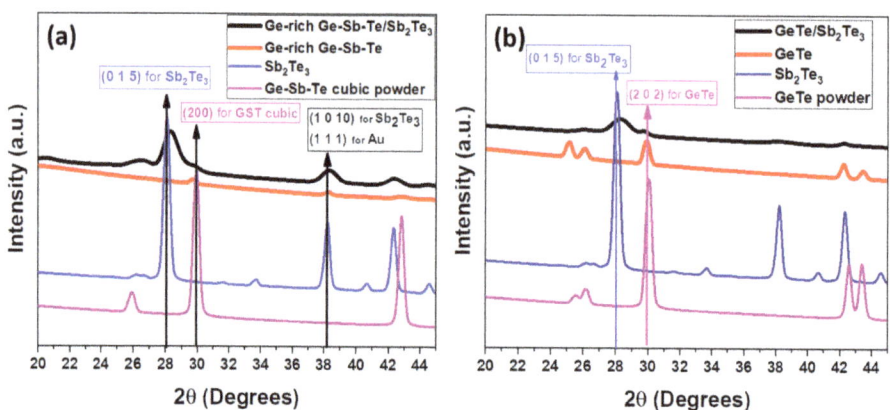

Figure 2. XRD patterns of (a) GGST, and GGST/ST core-shell NWs; (b) GT, and GT/ST core-shell NWs.

XPS analysis was employed to investigate the elemental composition and chemical bonding of the NWs in the form of core and core-shell nanostructures. In order to study the interfacial interaction between the core NWs and their corresponding shells, the XPS analysis of core-shell NWs was compared to the same analysis over the corresponding NWs without their shells. A detailed list of the samples analyzed and their related experimental XPS peak positions and FWHM are reported in the Supplementary Materials (Table S1).

The ex-situ XPS characterizations were performed straightaway after the NWs growth, to prevent the oxidation of the samples. The measurements were recorded with different take-off angles, i.e., the angle between the surface and detector. When samples are probed with a larger take-off angle, up to 90 degrees, the probed thickness through them is larger. By decreasing the take-off angle, shallower volumes are probed. Thus, upon changing the take-off angle, different volumes below the sample surface can be investigated, and this will be particularly relevant when different shell thicknesses are investigated.

If the relative intensities of the probed signals change by varying the take-off angles, we can predict that the signals are generated from our investigated material (NWs and Sb_2Te_3 thin film over the surface, as well as on the NWs) and not from the environmental contamination (such as carbon). The NWs grew uniformly over the surface, although along random orientations; most of them grew horizontally parallel to the surface, thus are suitable for XPS investigation. Exploiting the angle dependence of the measured depth, when no signal from the Ge atoms was detected, that were not expected in the shell, we were sure in particular to measure in the shell only.

As a first step, the sample containing GGST NWs catalyzed with Au NPs size of 50 nm size, was examined. The dots in the figures are the raw data, and the solid lines are the fitted data. In addition, in some figures, there are some lines of a gray color that show the individual deconvoluted peaks used for fitting. Figure 3a shows the XPS spectra of Ge, Sb, and Te obtained by analyses of the Ge_3d, Sb_4d, and Te_4d core level peaks. The gray color shows the inelastic background, fitted by the Shirley method, and the individual deconvoluted peaks used for fitting. The same data representation and analysis apply to Figure 4. Figure 3b shows the XPS spectrum of the Ge_2p core level, which has higher binding energy (1218 eV), and thus a smaller mean free path.

The inelastic mean free path (IMFP) is an index of how far an electron travels on average through a solid before losing energy, and can be calculated by the following formula:

$$I_{(d)} = I_0 \, e^{(-d/\lambda(E))} \tag{1}$$

where $I_{(d)}$ is the intensity after the primary electron beam has traveled through the solid to a distance d. I_0 is the intensity of the primary electrons, and λ is the interaction mean free path [33].

By considering IMFP for Ge_2p orbital electrons, which is about 0.9 nm, we can exclude that these electrons can probe an under layer buried by a 10 nm thick layer. On the contrary, if the shell thickness was not uniform, a Ge signal should emerge from the core, provided that, in some parts, the thickness is definitely thinner than 10 nm, comparable with the IMFP.

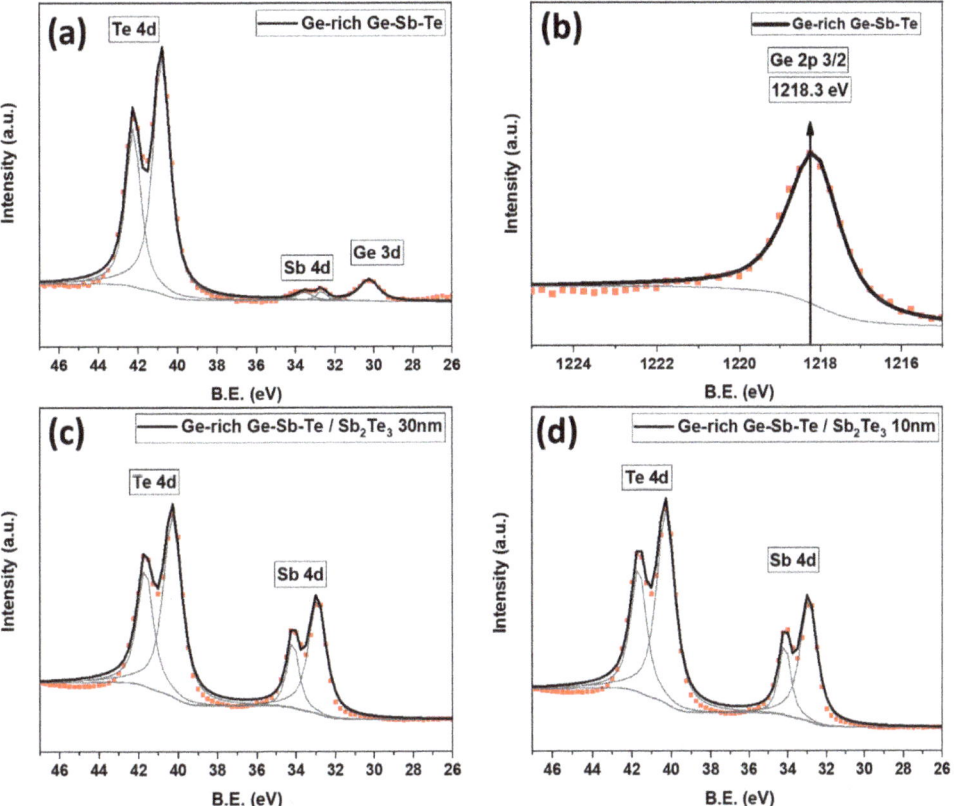

Figure 3. XPS spectra for (**a**) Te_4d, Sb_4d; and (**b**) Ge_2p region for GGST core NWs; (**c**) Te_4d, Sb_4d region for GGST/ST core-shell NWs with 30 nm shell thickness; (**d**) Te_4d, Sb_4d region for GGST/ST core-shell NWs with 10nm shell thickness; (dots–experimental data; solid lines—fit).

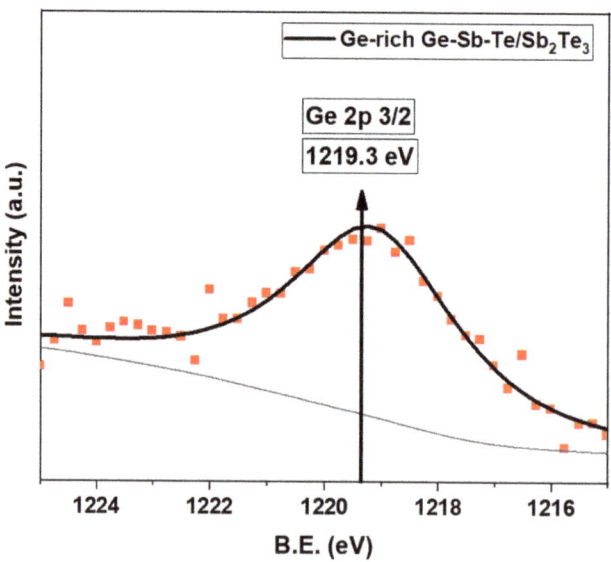

Figure 4. XPS spectra for Ge_2p$_{3/2}$ orbital of GGST/ST core-shell NWs with 10 nm shell thickness; (dots–experimental data; solid lines—fit).

In the Ge_2p range of energy, except Ge_2p, none of the NW shell (Te and Sb) XPS peaks present. So, we could study only the Ge signal from the NW core without any disturbance from the Sb$_2$Te$_3$ contributions, that could overlap with the peak, making the analysis more difficult and noisier. Ge is important for core and core-shell interface probing, because it is the only element present in the core of NWs but not in their shell. Thus, the Ge signal can give us information about the bulk and the interface of the NWs. The measurements confirmed the stability of Sb and Te [34]. The GGST experimental peak positions were clearly identified for each element and located at 30.20 ± 0.1 eV (Ge_3d), 1218.3 ± 0.1 eV (Ge_2p$_{3/2}$), 32.70 ± 0.1 eV (Sb_4d), and 40.15 ± 0.1 eV (Te_4d). The obtained spectrum confirmed the absence of oxidation on the NWs, as no extra peaks or shoulder tips related to oxidation were detected, within our experimental accuracy.

Next, we analyzed the GGST/ST core-shell NWs with different shell diameters. Figure 3c shows the XPS results on the NWs having 30 nm of shell thickness. The spectrum demonstrates the presence of Sb and Te elements only, with the peaks located at 32.70 ± 0.1 eV (Sb_4d) and 40.15 ± 0.1 eV (Te_4d), respectively. No presence of Ge was detected. This could be due to the fact that the ultimate depth that our XPS system could get data from is about 6 nm, this being a plausible reason for not observing the core contribution. Thus, in order to have a complete characterization of the core-shell structure and validate our results from an interdiffusion point of view, a lower shell thickness is required to probe the interface between the GGST and the ST shell. Moreover, it could be interesting to see whether Ge still remained unaffected as a core material after the shell's growth. Thus, we investigated the GGST/ST NWs with a conformal shell of 10 nm; Figure 3d displays the obtained XPS measurements. Only the presence of Sb and Te was detected, with peaks positions at 32.9 ± 0.1 eV (Sb_4d) and 40.2 ± 0.1 eV (Te_4d), respectively. Here, the absence of the Ge peak means that the core-shell interface is sharp and no Ge diffusion towards the shell takes place. A possible reason can be the room temperature shell deposition, that prevents interdiffusion. Thus, from the obtained results, we could confirm that the GGST core NWs are completely covered with the ST shell, and that Ge does not diffuse into the shell.

It should anyway be considered that all the surfaces of substrate and NWs are covered with Sb$_2$Te$_3$, whereas Ge falls only below the NWs' shell surface, and there is a low density of NWs compared to the substrate surface area. This makes the Sb and Te signals naturally larger than the Ge one, because they come from a wider region. Even when the few NWs are looked at, the Ge signal is attenuated by the shell thickness. The Sb_4d peak buries the Ge_3d (IMFP (λ), for Ge_3d is ~2 nm [35] in this energy region, so that the latter is hardly or not recognizable in the core-shell samples. For this reason, we considered a different energy region in which to study the Ge signal without any disturbance from other elements in the sample, such as Te and Sb. As mentioned before, (IMFP (λ) for Ge_2p is ~0.9 nm [36]. Thus, it is a suitable peak for looking at the Ge signal in core-shell NWs with 10 nm shell, as being isolated from other relevant peaks of ST.

Figure 4 shows the XPS spectra for Ge_2p$_{3/2}$ region of GGST core NWs and GGST/ST core-shell NWs with 10 nm shell thickness. The dots in the figures are the raw data, and the solid lines are the fitted data. The gray color shows the inelastic background, fitted by the Shirley method, and the individual deconvoluted peaks used for fitting. Thus, we were able to detect the Ge_2p peak in core-shell NWs with 10 nm shell thickness. Upon comparing such results with Figure 3b, an increase in the binding energy of the Ge_2p peak on the core-shell NWs was observed.

Moreover, looking at the increase in the binding energy of the Ge_2p peak, we concluded that the Ge atoms on the surface of the core GGST NWs tended to bond chemically with the Sb atoms from the shell, as the Ge-Sb bonding energy is more stable and has a low equilibrium potential. These results validated the existence of the chemical interaction between the GGST core and the ST shell. We noted that the same interaction appeared in the GT/ST case (data not shown).

We further analyzed the GGST/ST core-shell NWs having the shell thickness of 10 nm with different photoelectrons take-off angles. Figure 5a displays the fitted XPS spectra, with the same procedure as in Figure 3, featured by peaks at 32.70 ± 0.1 eV and 40.15 ± 0.1 eV, for Sb_4d and Te_4d, respectively. The increase in the peak intensity with the same peak position as a function of the incident angle was observed. However, the Ge peak was not observed. As discussed above, the very high peak intensity from the Te and Sb signals, which are related to the Sb_2Te_3 conformal coating, would overcome the Ge peak generated by the core NWs, having much less intensity. So, the Ge_2p peak at a higher binding energy was considered. Figure 5b shows the fitted XPS measurements of Ge_2p repeated with three different incident take-off angles, namely 30, 45, and 75 degrees. The Ge_2p core peak was observed for all angles at 1219.3 ± 0.1 eV (Ge_2p$_{3/2}$). By comparing the obtained results from different incident angles, an increase in peak intensity with the angle was observed, and it could demonstrate the existence of GGST as the core material in the core-shell NWs.

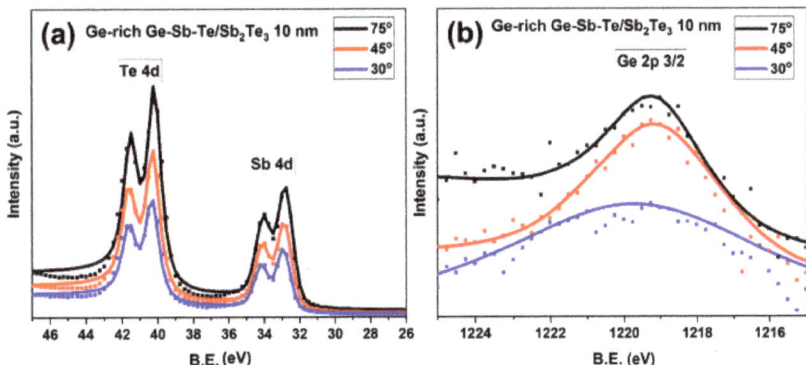

Figure 5. XPS spectra for (**a**) Te_4d, Sb_4d; and (**b**) Ge_2p regions collected at 30° (blue), 45° (red) and 75° (black) take-off angles for GGST/ST core-shell NWs with 10 nm shell thickness; (dots– experimental data; solid lines—fit).

Next, upon comparing the position of Ge_2p$_{3/2}$ in GGST/ST core-shell NWs (1219.3 eV) and GGST NWs (1218.3 eV, Figure 3b), a chemical shift in the binding energy of 1 eV was observed. Such a shift could be due to the chemical bond of GGST with ST.

Further, we performed XPS measurements on the GT core and GT/ST core-shell NWs. Figure 6a,b shows a clear and sharp presence of Ge_3d, Te_4d and Ge_2p$_{3/2}$ core peaks originating from the GT core NWs. The dots in the figures are the raw data, and the solid lines are the fitted data. The gray color shows the inelastic background, fitted by the Shirley method, and the individual deconvoluted peaks used for fitting. The same data representation and analysis apply to Figure 6c. Figure 6c reports the XPS analysis from the GT core NWs coated at room temperature with a 10 nm thick shell of ST. The spectrum shows clear Sb_4d and Te_4d peaks and an absence of the Ge_3d peak, suggesting the uniform coating of the ST shell over the GT core NWs. In addition, the Te_4d and Sb_4d peaks have exactly the same energy as in the GGST core-shell NWs with ST as the shell. It demonstrates that the shell has preserved its properties independently of the core material. The XPS analysis over the Ge_2p$_{3/2}$ peak was also acquired (see Supplementary Materials, Figure S1); the peak binding energy at 1218.5 eV is slightly increased in comparison with the value from the GT core NWs (1218.1 eV, Figure 6b), possibly claiming for the existence of a chemical interaction between the core and shell also for the GT/ST NWs.

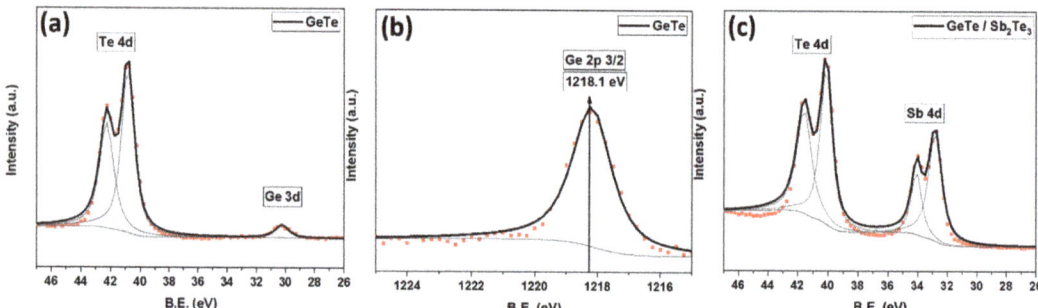

Figure 6. XPS spectra for (**a**) Ge_3d and Te_4d; (**b**) Ge_2p region for GT core NWs; (**c**) Sb_4d and Te_4d region for GT/ST core-shell with 10 nm shell thickness; (dots–experimental data; solid lines—fit).

Further, it is worth noting that the binding energy of Ge 2p$_{3/2}$ in GGST (1218.3 ± 0.1 eV) matches within the error bar, the value in GT (1218.1 ± 0.1 eV) meanwhile moves to higher energy in Ge$_2$Sb$_2$Te$_3$ (1219.5 ± 0.1 eV), as reported in the literature [37]. Therefore, the chemical bonding of Ge in GGST is more similar to the Ge in GT than to the Ge in Ge$_2$Sb$_2$Te$_5$.

However, after the growth of the shell, the Ge_2p peak binding energy is 1219.3 ± 0.1 eV in the GGST, being close to the value of 1219 ± 0.1 eV found in Ge$_2$Sb$_2$Te$_5$ and 1218.4 ± 0.1 eV in the GT cases. Such an observed trend may be related to the Ge atoms from the core NWs bonded with Sb atoms from the shell of NWs at the core–shell interface, supported by the tendency to form Ge–Sb bonding to reduce the overall potential energy of the alloy. When the value of Ge–Sb chemical bonding increases, the potential energy of the Ge atom decreases, however a detailed investigation on this mechanism is beyond the scope of this paper and would deserve more investigation.

4. Conclusions

In summary, Ge-rich Ge-Sb-Te and GeTe core NWs with a Sb$_2$Te$_3$ shell with thickness down to 10 nm were synthesized via MOCVD. The morphology showed a continuous shell coating all over the core NWs. The XRD analysis revealed that the core structure of the NWs was not altered by the shell deposition. The XPS measurements gave insight into the interaction between the NWs core and shell, with an indication about the Ge chemical state at the interface. The chemical shift of the Ge_2p peak was observed, confirming the interaction of the core and the shell. Angular-resolved XPS spectra indicated the absence of interdiffusion between the core and shell elements, suggesting that their structural phase can change independently, based on the alloy composition. This work demonstrated a straightforward method to provide efficient core-shell NW heterostructures, formed by two-phase change materials having different crystallization temperatures and reversible switching speed. This is particularly useful for comparison with corresponding planar multilayered PCM cells. Our results could be helpful in the fundamental understanding of phase change materials for the realization of memory devices and, in particular, for a comparison with corresponding planar multilayered PCM cells.

Supplementary Materials: The following supporting information can be downloaded at: https://www.mdpi.com/article/10.3390/nano12101623/s1, Table S1: Samples analyzed and their corresponding experimental XPS peak positions and FWHM; Figure S1: XPS spectra of Ge_2p region for GT/ST core-shell with 10 nm shell thickness; (dots–experimental data, solid lines–fit).

Author Contributions: Conceptualization, C.W.; methodology, A.K.; formal analysis, S.A.M.; investigation, A.L.; data curation, A.L.; writing—original draft preparation, A.K.; writing—review and editing, A.K., S.A.M., A.L.; M.C., M.L. visualization, A.K.; supervision, C.W.; funding acquisition, M.L. All authors have read and agreed to the published version of the manuscript.

Funding: This project has received funding from the European Union's Horizon 2020 Research and Innovation program under Grant Agreement No. 824957 ("BeforeHand: Boosting Performance of Phase Change Devices by Hetero- and Nanostructure Material Design").

Institutional Review Board Statement: Not applicable.

Informed Consent Statement: Not applicable.

Data Availability Statement: The data that support the findings of this study are available from the corresponding authors upon reasonable request.

Acknowledgments: We thank Raimondo Cecchini for his contribution to the development of NWs.

Conflicts of Interest: The authors declare no conflict of interest.

References

1. Ovshinsky, S.R. Reversible Electrical Switching Phenomena in Disordered Structures. *Phys. Rev. Lett.* **1968**, *21*, 1450. [CrossRef]
2. Lankhorst, M.H.R.; Ketelaars, B.W.S.M.M.; Wolters, R.A.M. Low-cost and nanoscale non-volatile memory concept for future silicon chips. *Nat. Mater.* **2005**, *4*, 347–352. [CrossRef] [PubMed]
3. Liu, Z.; Xu, J.; Chen, D.; Shen, G. Flexible electronics based on inorganic nanowires. *Chem. Soc. Rev.* **2014**, *44*, 161–192. [CrossRef] [PubMed]
4. Meyyappan, M.; Lee, J.S. The quiet revolution of inorganic nanowires. *IEEE Nanotechnol. Mag.* **2010**, *4*, 5–9. [CrossRef]
5. Eggleton, B.J.; Luther-Davies, B.; Richardson, K. Chalcogenide photonics. *Nat. Photonics* **2011**, *5*, 141–148. [CrossRef]
6. Yu, B.; Sun, X.; Ju, S.; Janes, D.B.; Meyyappan, M. Chalcogenide-nanowire-based phase change memory. *IEEE Trans. Nanotechnol.* **2008**, *7*, 496–502. [CrossRef]
7. Yamada, N.; Ohno, E.; Nishiuchi, K.; Akahira, N.; Takao, M. Rapid-phase transitions of GeTe-Sb$_2$Te$_3$ pseudobinary amorphous thin films for an optical disk memory. *J. Appl. Phys.* **1998**, *69*, 2849. [CrossRef]
8. Nishi, Y. *Advances in Non-Volatile Memory and Storage Technology*; Woodhead Publishing: Sawston, UK, 2014; ISBN 978-0-85709-803-0.
9. Lee, S.H.; Ko, D.K.; Jung, Y.; Agarwal, R. Size-dependent phase transition memory switching behavior and low writing currents in GeTe nanowires. *Appl. Phys. Lett.* **2006**, *89*, 223116. [CrossRef]
10. Rodgers, P.; Heath, J. *Nanoscience and Technology: A Collection of Reviews from Nature Journals*; World Scientific: Singapore, 2009; pp. 1–346. [CrossRef]
11. Cecchini, R.; Gajjela, R.S.R.; Martella, C.; Wiemer, C.; Lamperti, A.; Nasi, L.; Lazzarini, L.; Nobili, L.G.; Longo, M.; Cecchini, R.; et al. High-Density Sb2Te3 Nanopillars Arrays by Templated, Bottom-Up MOCVD Growth. *Small* **2019**, *15*, 1901743. [CrossRef]
12. Cecchini, R.; Selmo, S.; Wiemer, C.; Fanciulli, M.; Rotunno, E.; Lazzarini, L.; Rigato, M.; Pogany, D.; Lugstein, A.; Longo, M. In-doped Sb nanowires grown by MOCVD for high speed phase change memories. *Micro Nano Eng.* **2019**, *2*, 117–121. [CrossRef]
13. Cecchini, R.; Selmo, S.; Wiemer, C.; Rotunno, E.; Lazzarini, L.; De Luca, M.; Zardo, I.; Longo, M. Single-step Au-catalysed synthesis and microstructural characterization of core–shell Ge/In–Te nanowires by MOCVD. *Mater. Res. Lett.* **2017**, *6*, 29–35. [CrossRef]

14. Selmo, S.; Cecchini, R.; Cecchi, S.; Wiemer, C.; Fanciulli, M.; Rotunno, E.; Lazzarini, L.; Rigato, M.; Pogany, D.; Lugstein, A.; et al. Low power phase change memory switching of ultra-thin In3Sb1Te2 nanowires. *Appl. Phys. Lett.* **2016**, *109*, 213103. [CrossRef]
15. Longo, M. Advances in nanowire PCM. In *Advances in Non-Volatile Memory and Storage Technology*; Elsevier: Amsterdam, The Netherlands, 2019; pp. 443–518.
16. Wu, Y.; Fan, R.; Yang, P. Block-by-Block Growth of Single-Crystalline Si/SiGe Superlattice Nanowires. *Nano Lett.* **2002**, *2*, 83–86. [CrossRef]
17. Björk, M.T.; Ohlsson, B.J.; Sass, T.; Persson, A.I.; Thelander, C.; Magnusson, M.H.; Deppert, K.; Wallenberg, L.R.; Samuelson, L. One-dimensional heterostructures in semiconductor nanowhiskers. *Appl. Phys. Lett.* **2002**, *80*, 1058. [CrossRef]
18. Gudiksen, M.S.; Lauhon, L.J.; Wang, J.; Smith, D.C.; Lieber, C.M. Growth of nanowire superlattice structures for nanoscale photonics and electronics. *Nature* **2002**, *415*, 617–620. [CrossRef]
19. Lauhon, L.J.; Gudlksen, M.S.; Wang, D.; Lieber, C.M. Epitaxial core-shell and core-multishell nanowire heterostructures. *Nature* **2002**, *420*, 57–61. [CrossRef]
20. Dong, Y.; Yu, G.; McAlpine, M.C.; Lu, W.; Lieber, C.M. Si/a-Si core/shell nanowires as nonvolatile crossbar switches. *Nano Lett.* **2008**, *8*, 386–391. [CrossRef]
21. Jung, Y.; Ko, D.K.; Agarwal, R. Synthesis and structural characterization of single-crystalline branched nanowire heterostructures. *Nano Lett.* **2007**, *7*, 264–268. [CrossRef]
22. Yu, D.; Wu, J.; Gu, Q.; Park, H. Germanium telluride nanowires and nanohelices with memory-switching behavior. *J. Am. Chem. Soc.* **2006**, *128*, 8148–8149. [CrossRef]
23. Meister, S.; Peng, H.; McIlwrath, K.; Jarausch, K.; Zhang, X.F.; Cui, Y. Synthesis and characterization of phase-change nanowires. *Nano Lett.* **2006**, *6*, 1514–1517. [CrossRef]
24. Nukala, P.; Lin, C.C.; Composto, R.; Agarwal, R. Ultralow-power switching via defect engineering in germanium telluride phase-change memory devices. *Nat. Commun.* **2016**, *7*, 10482. [CrossRef] [PubMed]
25. Longo, M.; Wiemer, C.; Salicio, O.; Fanciulli, M.; Lazzarini, L.; Rotunno, E. Au-catalyzed self assembly of GeTe nanowires by MOCVD. *J. Cryst. Growth* **2011**, *315*, 152–156. [CrossRef]
26. Lee, S.H.; Jung, Y.; Agarwal, R. Highly scalable non-volatile and ultra-low-power phase-change nanowire memory. *Nat. Nanotechnol.* **2007**, *2*, 626–630. [CrossRef]
27. Longo, M.; Stoycheva, T.; Fallica, R.; Wiemer, C.; Lazzarini, L.; Rotunno, E. Au-catalyzed synthesis and characterisation of phase change Ge-doped Sb-Te nanowires by MOCVD. *J. Cryst. Growth* **2013**, *370*, 323–327. [CrossRef]
28. Jung, Y.; Lee, S.H.; Ko, D.K.; Agarwal, R. Synthesis and characterization of Ge2Sb2Te5 nanowires with memory switching effect. *J. Am. Chem. Soc.* **2006**, *128*, 14026–14027. [CrossRef]
29. Jung, Y.; Lee, S.H.; Jennings, A.T.; Agarwal, R. Core-shell heterostructured phase change nanowire multistate memory. *Nano Lett.* **2008**, *8*, 2056–2062. [CrossRef]
30. Kumar, A.; Cecchini, R.; Wiemer, C.; Mussi, V.; De Simone, S.; Calarco, R.; Scuderi, M.; Nicotra, G.; Longo, M. Phase Change Ge-Rich Ge–Sb–Te/Sb2Te3 Core-Shell Nanowires by Metal Organic Chemical Vapor Deposition. *Nanomaterials* **2021**, *11*, 3358. [CrossRef]
31. Kumar, A.; Cecchini, R.; Wiemer, C.; Mussi, V.; De Simone, S.; Calarco, R.; Scuderi, M.; Nicotra, G.; Longo, M. MOCVD Growth of GeTe/Sb2Te3 Core–Shell Nanowires. *Coatings* **2021**, *11*, 718. [CrossRef]
32. Available online: http://maud.radiographema.eu/ (accessed on 12 September 2021).
33. Egerton, R.F. *Electron Energy-Loss Spectroscopy in the Electron Microscope*; Springer: New York, NY, USA, 1996. [CrossRef]
34. Canvel, Y.; Lagrasta, S.; Boixaderas, C.; Barnola, S.; Mazel, Y.; Martinez, E. Study of Ge-rich GeSbTe etching process with different halogen plasmas. *J. Vac. Sci. Technol. A* **2019**, *37*, 031302. [CrossRef]
35. Shinotsuka, H.; Tanuma, S.; Powell, C.J.; Penn, D.R. Calculations of electron inelastic mean free paths. X. Data for 41 elemental solids over the 50 eV to 200 keV range with the relativistic full Penn algorithm. *Surf. Interface Anal.* **2015**, *47*, 871–888. [CrossRef]
36. Wang, W.; Lei, D.; Dong, Y.; Gong, X.; Tok, E.S.; Yeo, Y.C. Digital Etch Technique for Forming Ultra-Scaled Germanium-Tin ($Ge_{1-x}Sn_x$) Fin Structure. *Sci. Rep.* **2017**, *7*, 1835. [CrossRef] [PubMed]
37. Song, K.H.; Baek, S.C.; Lee, H.Y. Amorphous-to-crystalline phase transformation in $(GeTe)_x(Sb_2Te_3)$ (x = 0.5, 1, 2, 8) thin films. *J. Korean Phys. Soc.* **2012**, *61*, 10–16. [CrossRef]

MDPI
St. Alban-Anlage 66
4052 Basel
Switzerland
Tel. +41 61 683 77 34
Fax +41 61 302 89 18
www.mdpi.com

Nanomaterials Editorial Office
E-mail: nanomaterials@mdpi.com
www.mdpi.com/journal/nanomaterials

www.ingramcontent.com/pod-product-compliance
Lightning Source LLC
LaVergne TN
LVHW070618100526
838202LV00012B/674